Advance Praise for *Limits to Growth: The 30-Year Update*

"Thirty years have proven this model prophetic; now, in its newest iteration, we get one last challenge. May we pay more careful attention than in the past! We owe a great debt to the authors, including the late Donella Meadows, for whom this volume will serve as one of many fitting epitaphs."
—Bill McKibben, author, *The End of Nature*

"Reading the 30-year update reminds me of why the systems approach to thinking about our future is not only valuable, but indispensable. Thirty years ago, it was easy for the critics to dismiss the limits to growth. But in today's world, with its collapsing fisheries, shrinking forests, falling water tables, dying coral reefs, expanding deserts, eroding soils, rising temperatures, and disappearing species, it is not so easy to do so. We are all indebted to the *Limits* team for reminding us again that time is running out."
—Lester Brown, President, Earth Policy Institute

"Thirty years ago, *The Limits to Growth* was widely but erroneously attacked for prophesying doom, ignoring price, and denying adaptation. Today, with the global dynamics and challenges it foresaw now obvious to all, and the reforms it urged more vital than ever, its timely update remains an exceptionally valuable tool for understanding the unfolding future and creating the kind of future we want. Is there intelligent life on Earth? Work like this suggests grounds for cautious optimism."
—Amory B. Lovins, CEO, Rocky Mountain Institute

What the authors said in 1972:

"If the present growth trends in world population, industrialization, pollution, food production, and resource depletion continue unchanged, the limits to growth on this planet will be reached sometime within the next 100 years. The most probable result will be a rather sudden and uncontrolled decline in both population and industrial capacity."

How the critics responded:

"With current and near current technology, we can support 15 billion people in the world at twenty thousand dollars per capita for a millennium—and that seems to be a very conservative statement."

—Herman Kahn

"The material conditions of life will continue to get better for most people, in most countries, most of the time, indefinitely. Within a century or two, all nations and most of humanity will be at or above today's Western living standards."

—Julian Simon

The emerging consensus today:

"Human beings and the natural world are on a collision course. Human activities inflict harsh and often irreversible damage on the environment and on critical resources. If not checked, many of our current practices put at serious risk the future that we wish for human society and the plant and animal kingdoms, and may so alter the living world that it will be unable to sustain life in the manner that we know. Fundamental changes are urgent if we are to avoid the collision our present course will bring about."

—"World Scientists' Warning to Humanity"
*signed by more than 1,600 scientists, including
102 Nobel laureates, from 70 countries*

LIMITS TO GROWTH

LIMITS TO GROWTH

The 30-Year Update

DONELLA MEADOWS

JORGEN RANDERS

DENNIS MEADOWS

CHELSEA GREEN PUBLISHING COMPANY
WHITE RIVER JUNCTION, VERMONT

Printed in the United States of America
First printing, May, 2004

19 18 17 16   20 21 22 23

Printed on acid-free, recycled paper

Library of Congress Cataloging-in-Publication Data

Meadows, Donella H.
  Limits to growth : the 30-year update / Donella Meadows, Jorgen
Randers, and Dennis Meadows.
      p. cm.
  Includes bibliographical references and index.
  ISBN 1-931498-51-2 (hardcover : alk. paper) — ISBN 1-931498-58-X
(pbk. : alk. paper)
  1. Economic development—Environmental aspects. 2.
Population—Economic aspects. 3. Pollution—Economic aspects. 4.
Sustainable development. I. Randers, Jorgen. II. Meadows, Dennis L.
III. Title.
  HD75.6.M437 2004
  330.9—dc22

                              2004000125

Chelsea Green Publishing
85 North Main Street, Suite 120
White River Junction, VT 05001
(802) 295-6300
www.chelseagreen.com

Dedication

Over the past three decades many people and organizations have helped us understand how limits to material growth will shape global futures. We dedicate this volume to three individuals whose contributions were fundamental:

AURELIO PECCEI, founder of the Club of Rome, whose profound concern for the world and undying faith in humanity inspired us and many others to care about and address the prospects for humanity's long-term future.

JAY W. FORRESTER, professor emeritus of the Sloan School of Management at MIT and our teacher. He designed the prototype of the computer model we have used, and his profound systems insights have helped us understand the behaviors of economic and environmental systems.

Finally, it is our sad honor to dedicate this book to its main author, DONELLA H. MEADOWS. Widely known as Dana, by all those who respected her and appreciated her work, she was a world-class thinker, writer, and social innovator. Her high standards for communication, ethics, and service still inspire and challenge us—and thousands of others. Much of the analysis and prose here are hers, but this book was completed after Dana's death in February 2001. We intend that this edition will honor and advance her lifelong effort to inform the world's citizens and coax them toward sustainability.

Contents

Authors' Preface

Background

This book — *Limits to Growth: The 30-Year Update* — is the third edition in a series. The first text appeared in 1972.[1] In 1992 we published the revised edition, *Beyond the Limits (BTL)*,[2] where we discussed global developments over the first 20 years in the scenarios of *LTG*. This 30-year update presents the essential parts of our original analysis and summarizes some of the relevant data and the insights we have acquired over the past three decades.

The project that produced *LTG* took place in the System Dynamics Group of the Sloan School of Management within the Massachusetts Institute of Technology (MIT) from 1970 to 1972. Our project team used system dynamics theory and computer modeling to analyze the long-term causes and consequences of growth in the world's population and material economy. We addressed questions such as: *Are current policies leading to a sustainable future or to collapse? What can be done to create a human economy that provides sufficiently for all?*

We had been commissioned to examine these questions by the Club of Rome, an informal, international group of distinguished businessmen, statesmen, and scientists. The Volkswagen Foundation in Germany provided the funding for our work.

Dennis Meadows, then on the faculty at MIT, assembled and directed the following project team, which spent two years conducting the original study.

ALISON A. ANDERSON, PhD (USA)
ILYAS BAYAR (Turkey)
FARHAD HAKIMZADEH (Iran)
JUDITH A. MACHEN (USA)
DONELLA H. MEADOWS, PhD (USA)
NIRMALA S. MURTHY (India)
JORGEN RANDERS, PhD (Norway)
JOHN A. SEEGER, PhD (USA)

ERICH K.O. ZAHN, PhD (Germany)
JAY M. ANDERSON, PhD (USA)
WILLIAM W. BEHRENS III, PhD (USA)
STEFFEN HARBORDT, PhD (Germany)
PETER MILLING, PhD (Germany)
ROGER F. NAILL, PhD (USA)
STEPHEN SCHANTZIS (USA)
MARILYN WILLIAMS (USA)

A major foundation of our project was the "World3" computer model, which we constructed to help us integrate data and theories related to growth.[3] With the model we can produce scenarios of world development that are internally consistent. In the first edition of *LTG* we published and analyzed 12 scenarios from World3 that showed different possible patterns of world development over the two centuries from 1900 to 2100. *BTL* presented 14 scenarios from a slightly updated version of World3.

LTG became a best seller in several countries, eventually being translated into about 30 languages. *BTL* appeared in many languages and is widely used as a university text.

1972: The Limits to Growth

The Limits to Growth (LTG) reported that global ecological constraints (related to resource use and emissions) would have significant influence on global developments in the twenty-first century. *LTG* warned that humanity might have to divert much capital and manpower to battle these constraints—possibly so much that the average quality of life would decline sometime during the twenty-first century. Our book did not specify exactly what resource scarcity or what emission type might end growth by requiring more capital than was available—simply because such detailed predictions can not be made on a scientific basis in the huge and complex population–economy–environment system that constitutes our world.

LTG pleaded for profound, proactive, societal innovation through technological, cultural, and institutional change in order to avoid an increase in the ecological footprint of humanity beyond the carrying capacity of planet Earth. Although the global challenge was presented as grave, the tone of *LTG* was optimistic, stressing again and again how much one could reduce the damage caused by approaching (or exceeding) global ecological limits if early action were taken.

The 12 World3 scenarios in *LTG* illustrate how growth in population and natural resource use interacts with a variety of limits. In reality limits to growth appear in many forms. In our analysis we focused principally on the planet's physical limits, in the form of depletable natural resources and the finite capacity of the Earth to absorb emissions from industry and agriculture. In every realistic scenario we found that these limits force an end to physical growth in World3 sometime during the twenty-first century.

Our analysis did not foresee abrupt limits—absent one day, totally binding the next. In our scenarios the expansion of population and physical capital gradually forces humanity to divert more and more capital to cope with the problems arising from a combination of constraints. Eventually so much capital is diverted to solving these problems that it becomes impossible to sustain further growth in industrial output. When industry declines, society can no longer sustain greater and greater output in the other economic sectors: food, services, and other consumption. When those sectors quit growing, population growth also ceases.

The end to growth may take many forms. It can occur as a collapse: an uncontrolled decline in both population and human welfare. The scenarios of World3 portray such collapse from a variety of causes. The end to growth can also occur as a smooth adaptation of the human footprint to the carrying capacity of the globe. By specifying major changes in current policies we can cause World3 to generate scenarios with an orderly end to growth followed by a long period of relatively high human welfare.

The End of Growth

The end of growth, in whatever form, seemed to us to be a very distant prospect in 1972. All World3 scenarios in *LTG* showed growth in population and economy continuing well past the year 2000. Even in the most pessimistic *LTG* scenario the material standard of living kept increasing all the way to 2015. Thus *LTG* placed the end of growth almost 50 years after the publication of the book. That seemed to be time enough for deliberation, choice, and corrective action—even at the global level.

When we wrote *LTG* we hoped that such deliberation would lead society to take corrective actions to reduce the possibilities of collapse. Collapse is not an attractive future. The rapid decline of population and economy to levels that can be supported by the natural systems of the globe will no doubt be accompanied by failing health, conflict, ecological devastation, and gross inequalities. Uncontrolled collapse in the human footprint will come from rapid increases in mortality and rapid declines in consumption. With appropriate choice and action such uncontrolled decline could be avoided; overshoot could instead be resolved by a conscious effort to reduce humanity's demands on the planet. In this latter case gradual downward adjustment of the footprint would result from successful efforts to reduce fertility and from more equitable distribution of the sustainable rate of material consumption.

It is worth repeating that growth does not necessarily lead to collapse. Collapse follows growth only if the growth has led to overshoot, to an expansion in demands on the planet's sources, and sinks above levels that can be sustained. In 1972 it seemed that humanity's population and economy were still comfortably below the planet's carrying capacity. We thought there was still room to grow safely while examining longer-term options. That may have been true in 1972; by 1992 it was true no longer.

1992: Beyond the Limits

In 1992 we conducted a 20-year update of our original study and published the results in *Beyond the Limits*. In *BTL* we studied global developments between 1970 and 1990 and used this information to update the *LTG* and the World3 computer model. *BTL* repeated the original message; in 1992 we concluded that two decades of history mainly supported the conclusions we had advanced 20 years earlier. But the 1992 book did offer one major new finding. We suggested in *BTL* that humanity had already overshot the limits of Earth's support capacity. This fact was so important that we chose to reflect it in the title of the book.

Already in the early 1990s there was growing evidence that humanity was moving further into unsustainable territory. For example, it was reported that the rain forests were being cut at unsustainable rates; there was speculation that grain production could no longer keep up with population growth; some thought that the climate was warming; and there was concern about the recent appearance of a stratospheric ozone hole. But for most people this did not add up to proof that humanity had exceeded the carrying capacity of the global environment. We disagreed. In our view by the early 1990s overshoot could no longer be avoided through wise policy; it was already a reality. The main task had become to move the world back "down" into sustainable territory. Still, *BTL* retained an optimistic tone, demonstrating in numerous scenarios how much the damage from overshoot could be reduced through wise global policy, changes in technology and institutions, political goals, and personal aspirations.

BTL was published in 1992, the year of the global summit on environment and development in Rio de Janeiro. The advent of the summit seemed to prove that global society finally had decided to deal seriously with the important environmental problems. But we now know that humanity failed to achieve the goals of Rio. The Rio + 10 conference in Johannesburg in 2002 produced even less; it was almost paralyzed by a variety of ideological and economic disputes, by the efforts of those pursuing their narrow national, corporate, or individual self-interests.[4]

1970 – 2000: Growth in the Human Footprint

The past 30 years have produced many positive developments. In response to an ever growing human footprint, the world has implemented new technologies, consumers have altered their buying habits, new institutions have been created, and multinational agreements have been crafted. In some regions food, energy, and industrial production have grown at rates far exceeding population growth. In those regions most people have become wealthier. Population growth rates have declined in response to increased income levels. Awareness of environmental issues is much higher today than in 1970. There are ministries of environmental affairs in most countries, and environmental education is commonplace. Most pollution has been eliminated from the smoke stacks and outflow pipes of factories in the rich world, and leading firms are pushing successfully for ever higher eco-efficiency.

These apparent successes made it difficult to talk about problems of overshoot around 1990. The difficulty was increased by the lack of basic data and even elementary vocabulary related to overshoot. It took more than two decades before the conceptual framework—for example, distinguishing growth in the Gross Domestic Product (GDP) from growth in the ecological footprint—matured sufficiently to enable an intelligent conversation about the limits to growth issue. And world society is still trying to comprehend the concept of *sustainability*, a term that remains ambiguous and widely abused even sixteen years after the Brundtland Commission coined it.[5]

The past decade has produced much data that support our suggestion in *BTL* that the world is in overshoot mode. It now appears that the global per capita grain production peaked in the mid-1980s. The prospects for significant growth in the harvest of marine fish are gone. The costs of natural disasters are increasing, and there is growing intensity, even conflict, in efforts to allocate fresh water resources and fossil fuels among competing demands. The United States and other major nations continue to increase their greenhouse gas emissions even though scientific consensus and meteorological data both suggest that the global climate is being altered by human activity. There are already persistent economic declines in many localities and regions. Fifty-four nations, with 12 percent of the world population, experienced declines in per capita GDP for more than a decade during the period from 1990 to 2001.[6]

The past decade also provided new vocabulary and new quantitative measures for discussing overshoot. For example, Mathis Wackernagel and his colleagues measured the *ecological footprint* of humanity and compared it to the "carrying capacity" of the planet.[7] They defined the ecological footprint as the land area that would be required to provide the resources (grain, feed, wood, fish, and urban land) and absorb the emissions (carbon dioxide) of global society. When compared with the available land, Wackernagel concluded that human resource use is currently some 20 percent above the global carrying capacity (figure P-1). Measured this way humanity was last at sustainable levels in the 1980s. Now it has overshot by some 20 percent.

Sadly, the human ecological footprint is still increasing despite the progress made in technology and institutions. This is all the more serious because humanity is *already* in unsustainable territory. But the general awareness of this predicament is hopelessly limited. It will take a long time to obtain political support for the changes in individual values and public policy that could reverse current trends and bring the ecological footprint back below the long-term carrying capacity of the planet.

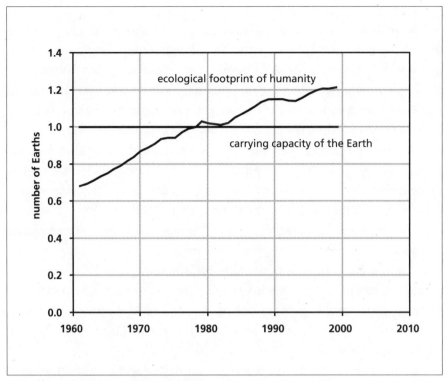

FIGURE P-1 Ecological Footprint versus Carrying Capacity
This graph shows the number of Earths required to provide the resources used by humanity and to absorb their emissions for each year since 1960. This human demand is compared with the available supply: our one planet Earth. Human demand exceeds nature's supply from the 1980s onward, overshooting it by some 20 percent in 1999. (Source: M. Wackernagel et al.)

What Will Happen?

The global challenge can be simply stated: To reach sustainability, humanity must increase the consumption levels of the world's poor, while at the same time reducing humanity's total ecological footprint. There must be technological advance, and personal change, and longer planning horizons. There must be greater respect, caring, and sharing across political boundaries. This will take decades to achieve even under the best of circumstances. No modern political party has garnered broad support for such a program, certainly not among the rich and powerful, who could make room for growth among the poor by reducing their own footprints. Meanwhile, the global footprint gets larger day by day.

Consequently, we are much more pessimistic about the global future than we were in 1972. It is a sad fact that humanity has largely squandered the past 30 years in futile debates and well-intentioned, but halfhearted, responses to the global ecological challenge. We do not have another 30 years to dither. Much will have to change if the ongoing overshoot is not to be followed by collapse during the twenty-first century.

We promised Dana Meadows before she died in early 2001 that we would complete the "30-year update" of the book she loved so much. But in the process we were once more reminded of the great differences among the hopes and expectations of the three authors.

Dana was the unceasing optimist. She was a caring, compassionate believer in humanity. She predicated her entire life's work on the assumption that if she put enough of the right information in people's hands, they would ultimately go for the wise, the farsighted, the humane solution—in this case, adopting the global policies that would avert overshoot (or, failing that, would ease the world back from the brink). Dana spent her life working for this ideal.

Jorgen is the cynic. He believes that humanity will pursue short-term goals of increased consumption, employment, and financial security to the bitter end, ignoring the increasingly clear and strong signals until it is too late. He is sad to think that society will voluntarily forsake the wonderful world that could have been.

Dennis sits in between. He believes actions will ultimately be taken to avoid the worst possibilities for global collapse. He expects that the world will eventually choose a relatively sustainable future, but only after severe global crises force belated action. And the results secured after long delay will be much less attractive than those that could have been attained through earlier action. Many of the planet's wonderful ecological treasures will be destroyed in the process; many attractive political and economic options will be lost; there will be great and persisting inequalities, increasing militarization of society, and widespread conflict.

It is impossible to combine these three outlooks into one common view of the most likely global future. But we do agree on what we hope might happen. The changes we would prefer to see are described in a slightly updated version of Dana's hopeful, concluding chapter from *BTL*, now titled "Tools for the Transition to Sustainability." The message is that if we persist in our pedagogic effort, the world's people will increasingly choose the right way ahead, out of love and respect for their planetary companions, current and future, human and nonhuman. We fervently hope they will do so in time.

Was Limits to Growth Correct?

We are often asked, "Were the *Limits to Growth* predictions correct?" Note that this is the media's language, not ours! We still see our research as an effort to identify different possible futures. We are not trying to predict the future. We are sketching alternative scenarios for humanity as we move toward 2100. Nonetheless it is useful to reflect on the lessons of the past 30 years. So, what has happened since *LTG* appeared as a slim paperback from an unknown publisher in Washington, DC, in March 1972?

At first the voices of most economists, along with many industrialists, politicians, and Third World advocates were raised in outrage at the idea of growth limits. But eventually events demonstrated that the concept of global ecological constraints is not absurd. There truly are limits to physical growth, and they have an enormous influence on the success of policies we choose to pursue our goals. And history does suggest that society has limited capacity for responding to those limits with wise, farsighted, and altruistic measures that disadvantage important players in the short term.

Resource and emission constraints have created many crises since 1972, exciting the media, attracting public attention, and arousing politicians. The decline in oil production within important nations, the thinning of stratospheric ozone, the mounting global temperature, the widespread persistence of hunger, the escalating debate over the location of disposal sites for toxic wastes, falling groundwater levels, disappearing species, and receding forests are just a few of the problems that have engendered major studies, international meetings, and global agreements. All of them illustrate and are consistent with our basic conclusion—that physical growth constraints are an important aspect of the global policy arena in the twenty-first century.

For those who respect numbers, we can report that the highly aggregated scenarios of World3 still appear, after 30 years, to be surprisingly accurate. The world in the year 2000 had the same number of people (about 6 billion—up from 3.9 billion in 1972) that we projected in the 1972 standard run of World3.[8] Furthermore, that scenario showed a growth in global food production (from 1.8 billion tons of grain equivalent per year in 1972 to 3 billion in 2000) that matches history quite well.[9] Does this correspondence with history prove that our model was true? No, of course not. But it does indicate that World3 was not totally absurd; its assumptions and our conclusions still warrant consideration today.

It is important to remember that one does not need to put World3 on a computer to understand its basic conclusions. Our most important statements about the likelihood of collapse do not come from blind faith in the curves generated by World3. They result simply from understanding the dynamic patterns of behavior that are produced by three obvious, persistent, and common features of the global system: erodable limits, incessant pursuit of growth, and delays in society's responses to approaching limits. Any system dominated by these features is prone to overshoot and collapse. The central assumptions of World3 consist of cause and effect mechanisms that produce limits, growth, and delays. Given that these mechanisms indisputably exist in the real world also, it should be no surprise that the world is evolving along a path that is consistent with the main features of the scenarios in *LTG*.

Why Another Book?

Why do we bother with publishing this updated version of *BTL*, if it is still making basically the same points as the two previous books? Our main goal is to restate our 1972 argument in a way that is more understandable and better supported by all the data and examples that have emerged during the past decades. In addition, we wish to give the many teachers who use our earlier texts updated materials for use with their students. *BTL* still gives useful views of the future, but it is a questionable practice for any teacher in the twenty-first century to assign a text with data tables ending in 1990.

And we have other reasons to write this text. We wish, once again, to

- stress that humanity is in overshoot and that the resulting damage and suffering can be greatly reduced through wise policy;
- offer data and analysis that contradict prevailing political pronouncements that humanity is on the correct path for its twenty-first century;
- inspire the world's citizens to think about the long-term consequences of their actions and choices—and muster their political support for actions that would reduce the damage from overshoot;
- bring the World3 computer model to the attention of a new generation of readers, students, and researchers;
- show what progress has been made since 1972 in understanding the long-term causes and consequences of growth.

Scenarios and Forecasting

We do *not* write this book in order to publish a forecast about what will actually happen in the twenty-first century. We are *not* predicting that a particular future will take place. We are simply presenting a range of alternative scenarios: literally, 10 different pictures of how the twenty-first century may evolve. We do this to encourage your learning, reflection, and personal choice.

We do not believe that available data and theories will ever permit accurate predictions of what will happen to the world over the coming century. But we do believe that current knowledge permits us to rule out a range of futures as unrealistic. Available facts already invalidate many people's implicit expectations of sustained growth in the future—they are just wishful thinking, attractive but erroneous, expedient but ineffective. Our analysis will be useful, if it forces citizens in the global society to reconsider and become more informed and respectful of the global physical limits that will play an important role in their future lives.

Books and the Transition to Sustainability

A book might seem like a weak tool in the struggle to attain sustainable development. But the history of our work gives a different view. Millions of copies of *LTG* and *BTL* were sold. The first book triggered a widespread debate, and the second one rekindled it. We did increase awareness and concern about environmental issues in the early days of the environmental movement. Many students who read *LTG* were led to adopt new career goals and to focus their studies on issues related to environment and sustainable development. That was all useful.

But our work fell short in many ways. The main ambition in *LTG* and *BTL* was to draw attention to the phenomenon of global ecological overshoot and to encourage society to question the pursuit of growth as a panacea for most problems. We did bring the phrase "limits to growth" into widespread use. The term, however, is often misunderstood, and it is typically used today in a very simplistic way. Most critics believe that our concerns about limits result from a belief that fossil fuels or some other resource will soon be exhausted. In fact our apprehension is more subtle; we worry that current policies will produce global overshoot and collapse through ineffective efforts to anticipate and cope with ecological limits. We believe that the human economy is exceeding important limits now and that this overshoot will intensify greatly over the coming decades. We failed in our earlier books to convey this concern in a lucid manner. We failed totally to get the concept of "overshoot" accepted as a legitimate concern for public debate.

It is useful to compare our results with those groups (largely comprised of economists) who have spent the past 30 years pushing the concept of free trade. Unlike us, they have been able to make their concept a household word. Unlike us, they have convinced numerous politicians to fight for free trade. But they, too, are faced with a widespread and fairly fundamental lack of conviction and fidelity that emerges whenever free trade policies also entail immediate personal or local costs, such as job losses. There are also many misconceptions about the total package of costs and benefits that result from adopting the goal of free trade. Ecological overshoot seems to us to be a much more important concept in the twenty-first century than free trade. But it is far behind in the fight for public attention and respect. This book is a new attempt to close that gap.

Overshoot and Collapse in Practice

Overshoot—and subsequent decline—in societal welfare will result when society does not prepare sufficiently well for the future. Welfare loss will occur, for example, when there is no ready replacement for dwindling reserves of oil, for scarcer wild fish, and for more expensive tropical woods, once these resources start to deplete. The problem is worse when the resource base is erodible and gets destroyed during overshoot. Then society might experience collapse.

One vivid example of global overshoot and collapse did actually take place around the turn of the millennium: the "dot.com bubble" in the global stock market. The bubble illustrates the dynamics of interest in this book, although in the world of finance and not in the world of physical resources. The erodible resource was investor confidence.

What happened, briefly, was that share prices rose spectacularly from 1992 to March 2000, to what was in retrospect a totally unsustainable peak. From this peak share values fell for a full three years before reaching a bottom in March 2003. Then prices gradually recovered (at least up to January 2004, when this is written).

Just as will be the case when humanity exceeds a resource or emission limit, there was little hardship associated with the long upturn in share prices. To the contrary, there was broad enthusiasm whenever share indices reached new heights. Most noteworthy, enthusiasm continued even after share prices had reached unsustainable territory—which in retrospect seems to have happened already in 1998. It was only long after the peak and some years into the collapse, that investors started to accept there had been a "bubble"—their word for overshoot. Once the collapse was well under way, no one could stop the fall. When it had lasted for three years, many doubted it would ever end. Investor confidence was completely eroded.

Sadly, we believe the world will experience overshoot and collapse in global resource use and emissions much the same was as the dot.com bubble—though on a much longer time scale. The growth phase will be welcomed and celebrated, even long after it has moved into unsustainable territory (this we know, because it has already happened). The collapse will arrive very suddenly, much to everyone's surprise. And once it has lasted for some years, it will become increasingly obvious that the situation before the collapse was totally unsustainable. After more years of decline, few will believe it will ever end. Few will believe that there once more will be abundant energy and sufficient wild fish. Hopefully they will be proved wrong.

Plans for the Future

Once the limits to growth were far in the future. Now they are widely in evidence. Once the concept of collapse was unthinkable. Now it has begun to enter into the public discourse—though still as a remote, hypothetical, and academic concept. We think it will take another decade before the consequences of overshoot are clearly observable and two decades before the fact of overshoot is generally acknowledged. The scenarios in this current volume show that the first decade of the twenty-first century will still be a period of growth—as did the scenarios in *LTG* 30 years ago. Our expectations for the 1970–2010 period therefore do not yet diverge much from those of our critics. We must all wait another decade for conclusive evidence about who has the better understanding.

We plan to update this report in 2012, on the fortieth anniversary of our first book. By then we expect there will be abundant data to test the reality of overshoot. We will be able to cite proof that we were right, or we will have to acknowledge data indicating that technology and the market have indeed boosted global limits far above the demands of human society. Population and economic declines will be imminent, or the world will be preparing for many more decades of growth. Until we can prepare that report, you will have to form your own opinion about causes and consequences of growth in the human ecological footprint. We hope you will find this compilation of information a useful basis for that effort.

January 2004
Dennis L. Meadows, Durham, N.H., USA
Jorgen Randers, Oslo, Norway

LIMITS TO GROWTH

CHAPTER 1

Overshoot

The future is no longer . . . what it might have been if humans had known how to use their brains and their opportunities more effectively. But the future can still become what we reasonably and realistically want.

—AURELIO PECCEI, 1981

To overshoot means to go too far, to go beyond limits accidentally—without intention. People experience overshoots every day. When you rise too quickly from a chair, you may momentarily lose your balance. If you turn on the hot-water faucet too far in the shower, you may be scalded. On an icy road your car might slide past a stop sign. At a party you may drink much more alcohol than your body can safely metabolize; in the morning you will have a ferocious headache. Construction companies periodically build more condominiums than are demanded, forcing them to sell units below cost and confront the possibility of bankruptcy. Too many fishing boats are often constructed. Then fishing fleets grow so large that they catch far more than the sustainable harvest. This depletes the fish population and forces ships to remain in harbor. Chemical companies have produced more chlorinated chemicals than the upper atmosphere can safely assimilate. Now the ozone layer will be dangerously depleted for decades until stratospheric chlorine levels decline.

The three causes of overshoot are always the same, at any scale from personal to planetary. First, there is growth, acceleration, rapid change. Second, there is some form of limit or barrier, beyond which the moving system may not safely go. Third, there is a delay or mistake in the perceptions and the responses that strive to keep the system within its limits. These three are necessary and sufficient to produce an overshoot.

Overshoot is common, and it exists in almost infinite forms. The change may be physical—growth in the use of petroleum. It may be organizational—an increase in the number of people supervised. It may be psychological—continuously rising goals for personal consumption. Or it may be manifest in financial, biological, political, or other forms.

The limits are similarly diverse—they may be imposed by a fixed amount of space; by limited time; by constraints inherent in physical, biological, political, psychological, or other features of a system.

The delays, too, arise in many ways. They may result from inattention, faulty data, delayed information, slow reflexes, a cumbersome or quarreling bureaucracy, a false theory about how the system responds, or from momentum that prevents the system from being stopped quickly despite the best efforts to halt it. For example, delays may result when a driver does not realize how much his car's braking traction has been reduced by ice on the road; the contractor uses current prices to make decisions about construction activity that will affect the market two or three years in the future; the fishing fleet owners base their decisions on data about recent catch, not information about the future rate of fish reproduction; chemicals require years to migrate from where they are used to a point in the ecosystem where cause severe damage.

Most instances of overshoot cause little harm. Being past many kinds of limits does not expose anyone to serious damage. Most types of overshoot occur frequently enough that when they are potentially dangerous, people learn to avoid them or to minimize their consequences. For example, you test the water temperature with your hand before stepping into the shower stall. Sometimes there is damage, but it is quickly corrected: Most people try to sleep extra long in the morning after a late night drinking in the bar.

Occasionally, however, there arises the potential for catastrophic overshoot. Growth in the globe's population and material economy confronts humanity with this possibility. It is the focus of this book.

Throughout this text we will grapple with the difficulties of understanding and describing the causes and consequences of a population and economy that have grown past the support capacities of the earth. The issues involved are complex. The relevant data are often poor in quality and incomplete. The available science has not yet produced consensus among researchers, much less among politicians. Nonetheless, we need a term that refers to the relation between humanity's demands on the planet and the globe's capacity to provide. For this purpose we will use the phrase *ecological footprint*.

The term was popularized by a study Mathis Wackernagel and his colleagues conducted for the Earth Council in 1997. Wackernagel calculated the amount of land that would be required to provide the natural resources consumed by the population of various nations and to absorb their wastes.[1] Wackernagel's term and mathematical approach were later adopted by the World Wide Fund for Nature (WWF), which provides semiannual data on the ecological footprint of more than 150 nations in its *Living Planet Report*.[2] According to these data, since the late 1980s the earth's peoples have been using more of the planet's resource production each year than could be regenerated in that year. In other words, the ecological footprint of global society has overshot the earth's capacity to provide. There is much information to support this conclusion. We will discuss it further in chapter 3.

The potential consequences of this overshoot are profoundly dangerous. The situation is unique; it confronts humanity with a variety of issues never before experienced by our species on a global scale. We lack the perspectives, the cultural norms, the habits, and the institutions required to cope. And the damage will, in many cases, take centuries or millennia to correct.

But the consequences need not be catastrophic. Overshoot can lead to two different outcomes. One is a crash of some kind. Another is a deliberate turnaround, a correction, a careful easing down. We explore these two possibilities as they apply to human society and the planet that supports it. We believe that a correction is possible and that it could lead to a desirable, sustainable, sufficient future for all the world's peoples. We also believe that if a profound correction is not made soon, a crash of some sort is certain. And it will occur within the lifetimes of many who are alive today.

These are enormous claims. How did we arrive at them? Over the past 30 years we have worked with many colleagues to understand the long-term causes and consequences of growth in human population and in its ecological footprint. We have approached these issues in four ways—in effect using four different lenses to focus on data in different ways, just as the lenses of a microscope and a telescope give different perspectives. Three of these viewing devices are widely used and easy to describe: (1) standard scientific and economic theories about the global system; (2) data on the world's resources and environment; and (3) a computer model to help us integrate that information and project its implications. Much of this book expands on those three lenses. It describes how we used them and what they allowed us to see.

Our fourth device is our "worldview," an internally consistent set of beliefs, attitudes, and values—a paradigm, a fundamental way of looking at reality. Everybody has a worldview; it influences where they look and what they see. It functions as a filter; it admits information consistent with their (often subconscious) expectations about the nature of the world; it leads them to disregard information that challenges or disconfirms those expectations. When people look out through a filter, such as a pane of colored glass, they usually see *through* it, rather than seeing *it*—and so, too, with worldviews. A worldview doesn't need to be described to people who already share it, and it is difficult to describe to people who don't. But it is crucial to remember that every book, every computer model, every public statement is shaped at least as much by the worldview of its authors as by any "objective" data or analysis.

We cannot avoid being influenced by our own worldview. But we can do our best to describe its essential features to our readers. Our worldview was formed by the Western industrial societies in which we grew up, by our scientific and economic training, and by lessons from traveling and working in many parts of the world. But the most important part of our worldview, the part that is least commonly shared, is our systems perspective.

Like any viewpoint—for example, the top of any hill—a systems perspective lets people see some things they would never have noticed from any other vantage point, and it may block the view of other things. Our training concentrated on dynamic systems—on sets of interconnected material and immaterial elements that change over time. Our training taught us to see the world as a set of unfolding behavior patterns, such as growth, decline, oscillation, overshoot. It has taught us to focus not so much on single pieces of a system as on connections. We see the many elements of demography, economy, and the environment as *one planetary system*, with innumerable interactions. We see stocks and flows and feedbacks and thresholds in the interconnections, all of which influence the way the system will behave in the future and influence the actions we might take to change its behavior.

The systems perspective is by no means the only useful way to see the world, but it is one we find particularly informative. It lets us approach problems in new ways and discover unsuspected options. We intend to share some of its concepts here, so you can see what we see and form your own conclusions about the state of the world and the choices for the future.

The structure of this book follows the logic of our global systems analysis. We have already made the basic point. Overshoot comes from the combination of (1) rapid change, (2) limits to that change, and (3) errors or delays in perceiving the limits and controlling the change. We will look at the global situation in that order: first at the driving factors that produce rapid global change, then at planetary limits, then at the processes through which human society learns about and responds to those limits.

We start in the next chapter with the phenomenon of change. Absolute, global rates of change are greater now than ever before in the history of our species. Such change is driven mainly by exponential growth in both population and the material economy. Growth has been the dominant behavior of the world socioeconomic system for more than 200 years. For example, figure 1-1 shows the growth of the human population, which is still surging upward despite dropping birth rates. Figure 1-2 shows that industrial output is growing, too, despite dips from oil price shocks, terrorism, epidemics, and other short-term influences. Industrial production has risen faster than population, resulting in an increase in the average material standard of living.

A consequence of growth in population and industry is change in many other features of the planetary system. For example, many pollution levels are growing. Figure 1-3 shows an important one, the accumulation in the atmosphere of carbon dioxide, a greenhouse gas, mainly as a result of fossil fuel burning and forest clearing by humans.

Other graphs throughout this book illustrate growth in food production, urban populations, energy consumption, materials use, and many other physical manifestations of human activity on the planet. Not everything is growing at the same rate or in the same way. As you can see from table 1-1, growth rates vary greatly. Some growth rates have come down, but they still produce significant annual increments in the underlying variable. Often a declining growth rate still produces a rising absolute increment, when a smaller percentage is multiplied by a much larger base. That is the case for 8 of the 14 factors in table 1-1. Over the past half century human

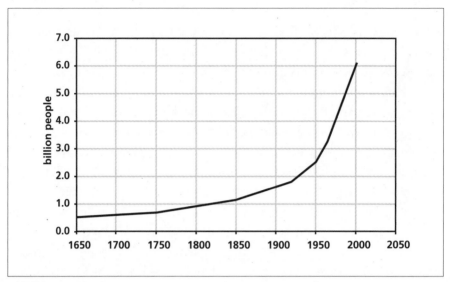

FIGURE 1-1 World Population

World population has been growing exponentially since the beginning of the Industrial Revolution. Note the shape of the curve and the increasing change as time goes on; these are hallmarks of exponential growth. The rate of growth is now falling, however; the curve is becoming less steep in a way that is just barely visible. In 2001 the world population growth rate was 1.3 percent per year, corresponding to a doubling time of 55 years. (Sources: PRB; UN; D. Bogue.)

beings have multiplied their own population, their physical possessions, and the material and energy flows they utilize by factors of 2, 4, 10, or even more, and they are hoping for more growth in the future.

Individuals support growth-oriented policies, because they believe growth will give them an ever increasing welfare. Governments seek growth as a remedy for just about every problem. In the rich world, growth is believed to be necessary for employment, upward mobility, and technical advance. In the poor world, growth seems to be the only way out of poverty. Many believe that growth is required to provide the resources necessary for protecting and improving the environment. Government and corporate leaders do all they can to produce more and more growth.

For these reasons growth has come to be viewed as a cause for celebration. Just consider some synonyms for that word: *development, progress, advance, gain, improvement, prosperity, success*.

Those are psychological and institutional reasons for growth. There are also what systems people call *structural* reasons, built into the connections among the elements of the population–economy system. Chapter 2 discusses

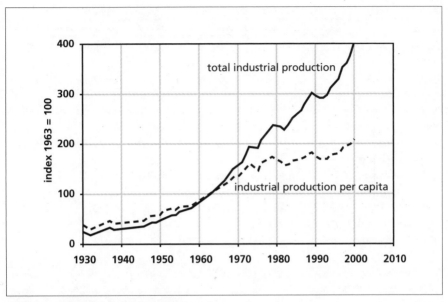

FIGURE 1-2 World Industrial Production

World industrial production, relative to the base year 1963, shows clear exponential increase, despite fluctuations due to oil price shocks and financial downturns. The growth rate over the past 25 years has averaged 2.9 percent per year, a doubling time of 25 years. The per capita growth rate has been slower, however, because of growth in population—only 1.3 percent per year, a doubling time of 55 years. (Sources: UN; PRB.)

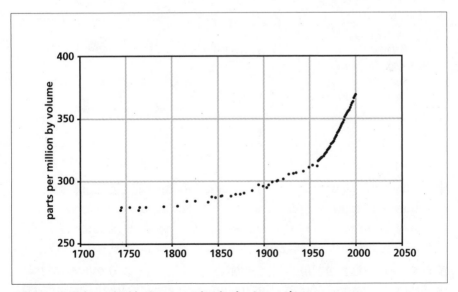

FIGURE 1-3 Carbon Dioxide Concentration in the Atmosphere

The concentration of carbon dioxide in the atmosphere has risen from roughly 270 parts per million (ppm) to more than 370 ppm and continues on its growth path. The sources of the carbon dioxide buildup are principally human fossil fuel burning and forest destruction. The consequence is global climate change. (Sources: UNEP; U.S. DoE.)

TABLE 1-1 Worldwide Growth in Selected Human Activities and Products 1950–2000

	1950	25-year change	1975	25-year change	2000
Human population (million)	2,520	160%	4,077	150%	6,067
Registered vehicles (million)	70	470%	328	220%	723
Oil consumption (million barrels per year)	3,800	540%	20,512	130%	27,635
Natural gas consumption (trillion cu. ft. per year)	6.5	680%	44.4	210%	94.5
Coal consumption (million metric tons per year)	1,400	230%	3,300	150%	5,100
Electrical generation capacity (million kilowatts)	154	1040%	1,606	200%	3,240
Corn (maize) production (million metric tons per year)	131	260%	342	170%	594
Wheat production (million metric tons per year)	143	250%	356	160%	584
Rice production (million metric tons per year)	150	240%	357	170%	598
Cotton production (million metric tons per year)	5.4	230%	12	150%	18
Wood pulp production (million metric tons per year)	12	830%	102	170%	171
Iron production (million metric tons per year)	134	350%	468	120%	580
Steel production (million metric tons per year)	185	350%	651	120%	788
Aluminum production (million metric tons per year)	1.5	800%	12	190%	23

(Sources: PRB; American Automobile Manufactures Association; Ward's Motor Vehicle Facts & Figures; U.S. DoE; UN; FAO; CRB.)

these structural causes of growth and describes their implications. There we will show why growth is such a dominant behavior of the world system.

Growth can solve some problems, but it creates others. That is because of limits, the subject of chapter 3. The Earth is finite. Growth of anything physical, including the human population and its cars and houses and factories, cannot continue forever. But the limits to growth are not limits to the number of people, cars, houses, or factories, at least not directly. They are limits to *throughput*—to the continuous flows of energy and materials needed to keep people, cars, houses, and factories functioning. They are limits to the rate at which humanity can extract resources (crops, grass, wood, fish) and emit wastes (greenhouse gases, toxic substances) without exceeding the productive or absorptive capacities of the world.

The population and economy depend upon air, water, food, materials, and fossil fuels from the earth. They emit wastes and pollution back to the earth. Sources include mineral deposits, aquifers, and the stock of nutrients in soils; among the sinks are the atmosphere, surface water bodies, and land-fills. The physical limits to growth are limits to the ability of planetary *sources* to provide materials and energy and to the ability of planetary *sinks* to absorb the pollution and waste.

In chapter 3 we examine the status of the earth's sources and sinks. The

data we present there make two points. One point is bad news; the other is good.

The bad news is that many crucial sources are emptying or degrading, and many sinks are filling up or overflowing. *The throughput flows presently generated by the human economy cannot be maintained at their current rates for very much longer.* Some sources and sinks are sufficiently stressed that they are already beginning to limit growth by, for instance, raising costs, increasing pollution burdens, and elevating the mortality rate.

The good news is that *current high rates of throughput are not necessary to support a decent standard of living for all the world's people.* The ecological foot-print could be reduced by lowering population, altering consumption norms, or implementing more resource-efficient technologies. These changes are possible. Humanity has the knowledge necessary to maintain adequate levels of final goods and services while reducing greatly the burden on the planet. In theory there are many possible ways to bring the human ecological footprint back down below its limits.

But theory does not automatically become practice. The changes and choices that will bring down the footprint are not being made, at least not fast enough to reduce the growing burden on the sources and sinks. They are not being made because there is no immediate pressure to make them, and because they take a long time to implement. That is the subject of chapter 4. There we discuss the signals that warn human society about the symptoms of its overshoot. And we examine the speed with which people and institutions can respond.

In chapter 4 we turn to our computer model, World3. It permits us to assemble many data and theories, putting the whole picture—growth, limits, response delays—into an explicit and coherent whole. And it gives us a tool for projecting the future consequences of our present understanding. We show what happens when the computer simulates the system as it might evolve, assuming no profound changes, no extraordinary efforts to see ahead, to improve signals, or to solve problems before they become critical.

The result of those simulations is, in nearly every scenario, overshoot and collapse of the planet's economy and population.

But not all scenarios show collapse. In chapter 5 we tell the best story we know about humanity's ability to look ahead, sense a limit, and pull back before experiencing disaster. We describe the international response to the news in the 1980s of a deteriorating stratospheric ozone layer. The story is important for two reasons. First, it provides a strong counterexample to the pervasive, cynical belief that people, governments, and corporations can never cooperate to solve global problems requiring foresight and self-discipline. Second, it illustrates concretely all three features required for overshoot: rapid growth, limits, and delayed response (in both science and politics).

The story of stratospheric ozone depletion and humanity's response now appears to be a success, but its final chapter won't be written for several more decades. So it is also a cautionary tale, an illustration of how perplexing it can be to guide the complex human enterprise toward sustainability within the interwoven systems of the planet while relying on imperfect understanding, delayed signals, and a system with enormous momentum.

In chapter 6 we use the computer for its primary purpose—not to predict what *will* result from current policies, but to ask what *could* happen if we make various changes. We build into the World3 model some hypotheses about human ingenuity. We concentrate on two mechanisms for problem solving—technology and markets—in which many people have placed great faith. Important features of those two remarkable human response capacities are already contained within World3, but in chapter 6 we strengthen them. We explore what would happen if the world society began to allocate its resources seriously to achieve pollution control, land preservation, human health, materials recycling, and much greater efficiency in the use of resources.

We discover from the resulting World3 scenarios that these measures help considerably. But by themselves they are not enough. They fall short, because technology–market responses are themselves delayed and imperfect. They take time, they demand capital, they require materials and energy flows, and they can be overwhelmed by population and economic growth. Technical progress and market flexibility will be necessary to avoid collapse and bring the world to sustainability. They are necessary, but they are not enough. Something more is required. That is the subject of chapter 7.

In chapter 7 we use World3 to explore what would happen if the industrial world were to supplement cleverness with wisdom. We assume the world adopts and begins to act upon two definitions of *enough*, one having to do with material consumption, the other with family size. These changes, combined with the technical changes we assumed in chapter 6, make possible a sustainable simulated world population of about eight billion. Those eight billion people all achieve a level of well-being roughly equivalent to the lower-income nations of present-day Europe. Given reasonable assumptions about market efficiency and technical advance, the material and energy throughputs needed by that simulated world could be maintained by the planet indefinitely. We show in this chapter that overshoot can ease back down to sustainability.

Sustainability is a concept so foreign to our present growth-obsessed culture that we take some time in chapter 7 to define it and to outline what a sustainable world might be like—and what it need *not* be like. We see no reason why a sustainable world needs to leave anyone living in poverty. Quite the contrary, we think such a world would have to provide material security to all its people. We don't think a sustainable society need be stagnant, boring, uniform, or rigid. It need not be, and probably could not be, centrally controlled or authoritarian. It could be a world that has the time, the resources, and the will to correct its mistakes, to innovate, to preserve the fertility of its planetary ecosystems. It could focus on mindfully increasing the quality of life rather than on mindlessly expanding material consumption and the physical capital stock.

The concluding chapter 8 derives more from our mental models than from data or a computer model. It gives the results of our personal attempts to understand what must be done now. Our world model, World3, gives the basis both for pessimism and for optimism about the future. And on this issue, the authors diverge. Dennis and Jorgen have come to believe that a decline in the average quality of life is now inevitable, and probably even global population and economy will be forced to fall. Donella believed all her life that humanity will develop the insights, institutions, and ethics it needs to achieve an attractive, sustainable society. But even with our different views we all three agreed on how the challenge should be approached, and this is discussed in chapter 8.

The first section of our final chapter lays out the priorities for action that could minimize the damage done to the planet and to society. The second section describes five tools that can help global society move toward a sustainable state.

Whatever lies ahead, we know its main dimensions will emerge over the next two decades. The global economy is already so far above sustainable levels that there is very little time left for the fantasy of an infinite globe. We know that adjustment will be a huge task. It will entail a revolution as profound as the agricultural and industrial revolutions. We appreciate the difficulty of finding solutions to problems such as poverty and employment, for which growth has been, so far, the world's only widely accepted hope. But we also know that reliance on growth involves a false hope, because such growth cannot be sustained. Blind pursuit of physical growth in a finite world ultimately makes most problems worse; better solutions to our real problems are possible.

Much that we wrote in *The Limits to Growth* 30 years ago remains true. But science and society have evolved over the past three decades. All of us have learned much and gained new perspectives. The data, the computer, and our own experience all tell us that the possible paths into the future have narrowed since we first addressed limits to growth in 1972. Levels of affluence we might have provided sustainably to all the globe's people are no longer attainable; ecosystems we might have preserved have been extinguished; resources that might have given wealth to future generations have been consumed. But there are still many available choices, and they are crucial. Figure 1-4 illustrates the enormous range of possibilities we believe still exists. The figure was derived by superimposing the curves for human population and human welfare generated by the 9 relevant computer scenarios we present later in this book.[3]

The set of possible futures includes a great variety of paths. There may be abrupt collapse; it is also possible there may be a smooth transition to sustainability. But the possible futures do not include indefinite growth in physical throughput. That is not an option on a finite planet. The only real choices are to bring the throughputs that support human activities down to sustainable levels through human choice, human technology, and human organization, or to let nature force the decision through lack of food, energy, or materials, or through an increasingly unhealthy environment.

In 1972 we opened *The Limits to Growth* with a quotation from U Thant, who was then secretary-general of the United Nations:

> I do not wish to seem overdramatic, but I can only conclude from the information that is available to me as Secretary-General, that the Members of the United Nations have perhaps ten years left in which to subordinate their ancient quarrels and launch a global partnership to curb the arms race, to improve the human environment, to defuse the population explosion, and to supply the required momentum to development efforts. If such a global partnership is not forged within the next decade, then I very much fear that the problems I have mentioned will have reached such staggering proportions that they will be beyond our capacity to control.[4]

More than 30 years have passed, and the global partnership is still not in

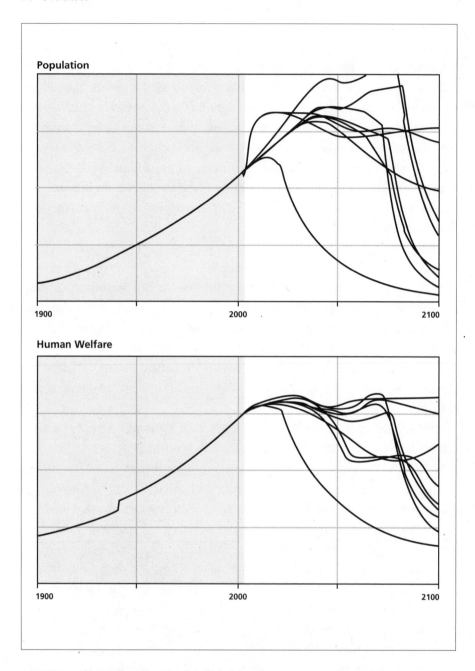

FIGURE 1-4 Alternative Scenarios for Global Population and Human Welfare
This figure superimposes all relevant World3 scenarios shown in this book to illustrate the wide range of possible paths for two important variables—population and average human welfare (measured as an index combining per capita income with other indicators of well-being). Most scenarios show decline, but some reflect a society that achieves a stable population and high, sustainable human welfare.

evidence. But there is growing consensus that humanity is immersed in problems beyond its control. And a great deal of data and many new studies support the secretary-general's warning.

For example, U Thant's concerns were echoed in a 1992 report, "World Scientists' Warning to Humanity" signed by more than 1,600 scientists, including 102 Nobel laureates, from 70 countries:

> Human beings and the natural world are on a collision course. Human activities inflict harsh and often irreversible damage on the environment and on critical resources. If not checked, many of our current practices put at serious risk the future that we wish for human society and the plant and animal kingdoms, and may so alter the living world that it will be unable to sustain life in the manner that we know. Fundamental changes are urgent if we are to avoid the collision our present course will bring about.[5]

The warning was even supported by a 2001 report authored within the World Bank:

> . . . an alarming rate of environmental degradation has occurred and in some cases is accelerating. . . . Across the developing world, environmental problems are imposing severe human, economic, and social costs and threatening the foundation upon which growth and, ultimately, survival depend.[6]

Was U Thant right? Are the world's current problems already beyond anyone's capacity to control? Or was he premature; might the confident statement of the 1987 World Commission on Environment and Development be correct?

> Humanity has the ability to make development sustainable—to ensure that it meets the needs of the present without compromising the ability of future generations to meet their own needs.[7]

No one can answer those questions for you with complete assurance. Yet it is urgently important that everyone develop well-considered answers to the questions above. Those answers are required for interpreting unfolding events and guiding personal actions and choices day by day.

We invite you to accompany us through the following discussion of the data, the analyses, and the insights we have accumulated over the past 30 years. And then you will have a basis for reaching your own conclusions about global futures and for making the choices that can guide your own life.

CHAPTER 2

The Driving Force: Exponential Growth

I find to my personal horror that I have not been immune to naïveté about exponential functions. . . . While I have been aware that the interlinked problems of loss of biological diversity, tropical deforestation, forest dieback in the northern hemisphere and climate change are growing exponentially, it is only this very year that I think I have truly internalized how rapid their accelerating threat really is.

—Thomas E. Lovejoy, 1988

The first cause of overshoot is growth, acceleration, rapid change. For more than a century many physical features of the global system have been growing rapidly. For example, population, food production, industrial production, consumption of resources, and pollution are all growing, often more and more rapidly. Their increase follows a pattern that mathematicians call *exponential growth*.

This pattern is extremely common. Figures 2-1 and 2-2 illustrate two very different examples, the tons of soybean produced each year and the number of people in less developed regions living in urban areas. Weather extremes, economic fluctuations, technical change, epidemics, or civil disruption may impose small ups and downs on the smooth curves, but on the whole exponential growth has been a dominant behavior of the human socioeconomic system since the industrial revolution.

This type of growth has surprising characteristics that make it very hard to manage. We will therefore preface our analysis of long-term options by defining exponential growth, describing its causes, and discussing the factors that govern its course. Physical growth on a finite planet must eventually end. But when will it end; what forces will cause it to decline? In what condition will humanity and the global ecosystem be left after it has ceased? To answer those questions it is necessary to understand the system structure

FIGURE 2-1 World Soybean Production
World soybean production has been increasing since 1950 with a doubling time of 16 years. (Sources: Worldwatch Institute; FAO.)

that makes the human population and economy constantly strive toward growth. That system is at the core of the World3 model, and, we believe, it is a defining feature of global society.

The Mathematics of Exponential Growth

Take a large piece of cloth and fold it in half. You've just doubled its thickness. Fold it in half again to make it four times as thick. Fold it in half again. Fold it in half a fourth time. Now it is 16 times as thick as the original—about a centimeter, or 0.4 inch, thick.

If you could go on folding the cloth that way 29 more times for a total of 33 doublings, how thick do you think it would become? Less than a foot? Between 1 foot and 10 feet? Between 10 feet and a mile?

Of course you can't fold a piece of cloth in half 33 times. But if you could, the bundle of cloth would be long enough to reach from Boston to Frankfurt—3,400 miles, about 5,400 kilometers.[1]

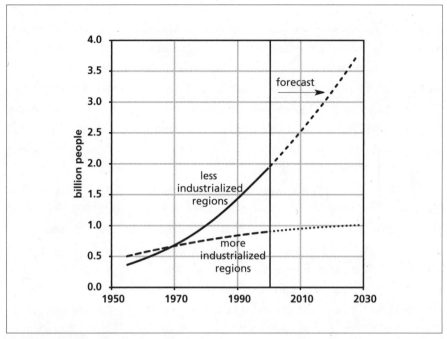

FIGURE 2-2 World Urban Population
Over the past half century the urban population has increased exponentially in the less industrialized regions of the world but almost linearly in the more industrialized regions. Average doubling time for city populations in less industrialized regions has been 19 years. This trend is expected to continue for several decades. (Source: UN.)

Exponential growth—the process of doubling and redoubling and redoubling again—is surprising, because it produces such huge numbers so quickly. Exponentially growing quantities fool us because most of us think of growth as a linear process. A quantity grows *linearly* when *its increase is a constant amount over a given period of time.* If a construction crew produces a mile of highway each week, the road grows linearly. If a child puts $7 each year in a jar, the savings increase linearly. The amount of new asphalt added is not affected by the length of the road already built, nor is the amount of annual savings affected by the money already in the jar. When some factor experiences linear growth, *the amount of its increase is always the same in a given time period;* it does not depend on how much of the factor has already accumulated.

A quantity grows *exponentially* when *its increase is proportional to what is already there.* A colony of yeast cells in which each cell divides into two every

10 minutes is growing exponentially. For each single cell, after 10 minutes there will be two cells. After the next 10 minutes there will be 4 cells, 10 minutes later there will be 8, then 16, and so on. The more yeast cells there are, the more new ones are made per unit of time. A company that successfully increases its gross sales by some percentage year after year will grow exponentially. *When some factor experiences exponential growth, the amount of its increase rises from one period to the next*; it depends on how much of the factor has already accumulated.

The vast difference between linear growth and exponential growth is illustrated by considering two ways to increase the sum of $100—you could put the money in a bank account to accumulate the interest, or you could put the money in a jar and add a fixed amount to it annually. If you put a single deposit of $100 in a bank that pays 7 percent interest per year, compounded annually, and let the interest income accumulate in the account, the invested money will grow exponentially. Every year there will be an addition to the money already there. The rate of the addition is constant at 7 percent per year, but the absolute amount of increase will grow. This addition is $7 at the end of the first year. The second year's interest will be 7 percent of $107, which is $7.49, bringing the total to $114.49 at the start of year three. One year later the interest will be $8.01, and the total will be $122.50. By the end of the 10th year the account will have grown to $196.72.

If you put $100 in a jar and add $7 to the contents each year, the money will grow linearly. At the end of the first year the jar will also hold $107, the same as the bank account. At the end of the 10th year it will hold $170, less money than there is in the bank account, but not a whole lot less.

Initially both saving strategies seem to generate quite similar results, but the explosive effect of sustained exponential accumulation eventually becomes apparent (figure 2-3). After the 20th year, the jar holds $240, while the bank account holds almost $400. By the end of the 30th year, linear growth in the jar will have produced $310 in savings. The bank account, with 7 percent annual interest, will stand at just over $761. So in 30 years, exponential growth at 7 percent per year produces more than two times as much as linear growth, even though both started with identical deposits. At the end of year 50, the bank account is 6.5 times bigger than the deposit in the jar—almost $2,500 more!

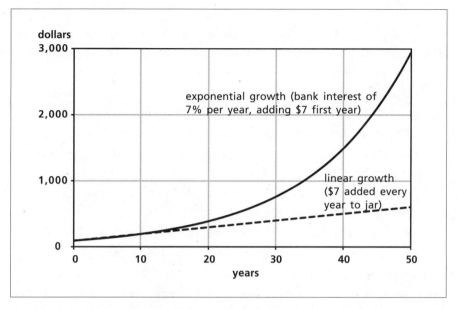

FIGURE 2-3 Linear versus Exponential Growth of Savings
If a person puts $100 in a jar and each year adds $7, the savings will grow linearly, as shown by the dashed line. If the person invests $100 in a bank at 7 percent per year interest, that $100 will grow exponentially, with a doubling time of about 10 years.

The unexpected consequences of exponential growth have fascinated people for centuries. A Persian legend tells about a clever courtier who presented a beautiful chessboard to his king and requested that the king give him in exchange one grain of rice for the first square on the board, two grains for the second square, four grains for the third, and so forth.

The king agreed and ordered rice to be brought from his stores. The fourth square on the chessboard required 8 grains, the tenth square took 512 grains, the fifteenth required 16,384, and the twenty-first square gave the courtier more than a million grains of rice. By the forty-first square, a trillion (10^{12}) rice grains had to be provided. The payment could never have continued to the sixty-fourth square; it would have taken more rice than there was in the whole world!

A French riddle illustrates another aspect of exponential growth— the apparent suddenness with which an exponentially growing quantity approaches a fixed limit. Suppose you own a pond. One day you notice that a single water lily is growing on your pond. You know that the lily plant will double in size each day. You realize that if the plant were allowed to grow unchecked, it would completely cover the pond in 30 days, choking off the other forms of life in the water. But initially the lily seems small, so you decide not to worry. You'll deal with it when it covers half the pond. How much time have you given yourself to prevent the destruction of your pond?

You have left yourself just one day! On the 29th day the pond is half covered. The next day—after one final doubling—the pond will be totally shaded. It initially seems reasonable to postpone action until the pond is half covered. On the 21st day, the plant covers just 0.2 percent of the pond. On the 25th, the plant covers just 3 percent of the pond. But again, that policy allows just one day to save your pond.[2]

You can see how exponential growth, combined with response delays, can lead to overshoot. For a long time the growth looks insignificant. There appears to be no problem. Then suddenly change comes on faster and faster, until, with the last doubling or two, there is no time to react. The apparent crisis of the lily pond's last day does not come from any change in the underlying process; the lily's percentage growth rate remains absolutely constant throughout the month. Still, that exponential growth accumulates suddenly to produce a problem that is unmanageable.

You could personally experience this sudden shift from insignificance to overload. Imagine eating one peanut on the first day of the month, two peanuts on the second day, four peanuts on the third, and so on. Initially you are buying and consuming an insignificant amount of food. But long before the end of the month, your bank account and your health would be severely affected. How long could you maintain this exercise in exponentially growing food intake, with a doubling time of one day? On the 10th day you would need to consume less than a pound of peanuts. But on the last day of the month, your policy of doubling consumption on each successive day would force you to buy and eat more than 500 tons of peanuts!

The peanut experiment would not cause serious harm, because you would simply, one day, contemplate an impossibly large pile of peanuts and quit. In this example there are no significant delays between when you take an action and when you feel its full consequences.

A quantity growing according to a pure exponential growth equation doubles in a constant time period. For the yeast colony, the doubling time was 10 minutes. Money in a bank earning 7 percent annual interest doubles about every 10 years. For the lily plant and the peanut experiment, the doubling times were each precisely one day. There is a simple relationship between the rate of growth in percentage terms and the time it will take a quantity to double. The doubling time is approximately equal to 72 divided by the growth rate in percent.[3] This is illustrated in table 2-1.

TABLE 2-1 Doubling Times

Growth Rate (% per year)	Approximate Doubling Times (years)
0.1	720
0.5	144
1.0	72
2.0	36
3.0	24
4.0	18
5.0	14
6.0	12
7.0	10
10.0	7

We can use the example of Nigeria to illustrate the consequences of sustained doubling. Nigeria had a population in 1950 of about 36 million. In the year 2000 its population was about 125 million. Over the second 50 years of the 20th century, Nigeria's population grew nearly fourfold. In the year 2000 its growth rate was reported to be 2.5 percent per year.[4] The corresponding doubling time was about 72 divided by 2.5 or approximately 29 years. If this population growth rate continued unchanged into the future, Nigeria's population would follow a path like the one illustrated in table 2-2.

A Nigerian child born in the year 2000 entered a population four times larger than Nigeria's population was in 1950. If the country's growth

TABLE 2-2 Nigeria's Population Growth, Extrapolated

Year	Population (million people)
2000	125
2029	250
2058	500
2087	1000

remains constant after the year 2000, and that child lives for 87 years, she will see the population multiply *another eightfold*. Late in the twenty-first century, there would be 8 Nigerians for every 1 in 2000, and 28 for every 1 in 1950. More than one billion people would live in Nigeria!

Nigeria is already one of many countries experiencing hunger and environmental deterioration. Clearly its population cannot expand another 8 times! The only reason for doing a calculation like the one in table 2-2 is to illustrate the algebra of doubling times and to demonstrate that *exponential growth never can go on very long in a finite space with finite resources*.

Why, then, is this growth going on in the world now? And what is likely to stop it?

Things That Grow Exponentially

Exponential growth occurs in two different ways. If an entity is self-reproducing, then its exponential growth is inherent. If an entity is driven by something else that is growing exponentially, then its growth is derived.

All living creatures, from bacteria to people, fall under the first category. Creatures are produced by creatures. We illustrate the system structure of a self-reproducing population with a diagram like this:

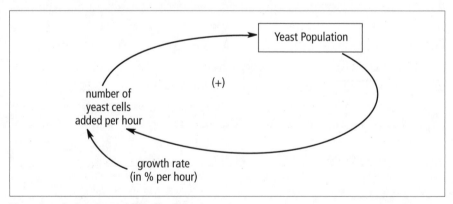

Yeast Population Growth Feedback Loop

The format of the diagram above is taken from our discipline, system dynamics, and it is quite precise. The box around the yeast population indicates that it is a *stock*—an accumulation, the net result of all past processes to increase and decrease the yeast. The arrows indicate causation or influence, which may be exerted in many ways. In this diagram the top arrow represents the influence of physical flows; it means that new yeast cells flow into and increase the stock, the yeast population. The bottom arrow represents the influence of information; it means that the size of the stock affects the production of new yeast. The greater the stock, the more new cells can be produced, as long as nothing changes the growth rate. (Things do change the growth rate, of course. For simplicity, they are omitted from this diagram. We'll come to them later.)

The (+) sign in the middle of the loop means that the two arrows together make up a *positive* or *reinforcing feedback loop*. A positive feedback loop is a chain of cause-and-effect relationships that closes on itself to create self-reinforcing change. It operates so that a change to any element anywhere in the loop will have consequences that cascade along the chain of causal links, finally changing the original element even more in the same direction. An increase will cause further increase; a decrease will eventually cause further decrease.

In system dynamics the title *positive loop* does not necessarily mean that the loop produces favorable results. It simply refers to the *reinforcing* direction of the causal influence around the loop. Similarly, negative feedback loops, which we will discuss further in a moment, do not necessarily produce unfavorable results. In fact they're often stabilizing. They are negative in the sense that they *counteract* or *reverse* or *balance* causal influence around the loop.

A positive feedback loop can operate as a "virtuous circle," or a "vicious circle," depending on whether the growth it produces is wanted or not. Positive feedback causes the exponential growth of yeast in rising bread and of money in your interest-bearing bank account. Those are useful. Positive feedback may also be responsible for pest outbreaks in an agricultural crop or for growth of a cold virus in your throat. Those are not useful.

Whenever a system stock is embedded in a positive feedback loop, that stock has the *potential* to grow exponentially. That doesn't mean it *will* grow exponentially; it does, however, have the *capacity* to do so if it is freed from constraints. Growth can be constrained by many things, such as lack of nutrients (in the case of yeast), low temperature and the presence of other populations (in the case of pests), and, in the case of the human population, incentives, disincentives, goals, purposes, disasters, diseases, desires. The rate of growth will vary over time; it will differ from place to place. But yeast, pest, or population growth, when it is not limited by a constraint, occurs exponentially.

The stock of *industrial capital* is something else that can exhibit inherent exponential growth. Machines and factories collectively can make other machines and factories. A steel mill can make the steel for another steel mill; a nuts-and-bolts factory can make nuts and bolts that hold together machines that make nuts and bolts; any business that makes a profit generates money for investment to expand the business. Both physical and monetary capital make even more capital possible, in the self-reproducing, growth-oriented fashion of the industrial economy.

It is not an accident that the industrial world has come to expect an economy to grow by a certain percentage of itself—say, 3 percent—each year. That expectation evolved out of several centuries of experience with capital creating more capital. It has become customary to save and invest for the future, to set aside a certain fraction of total output in the expectation that this will be invested to generate even more output in the future. An economy will grow exponentially whenever the self-reproduction of capital is unconstrained by consumer demand, labor availability, raw materials, energy, investment funds, or any of the other factors that can limit the growth of a complex industrial system. Like population, capital has the inherent system *structure* (a positive feedback loop) to produce the *behavior* of exponential growth. Economies don't always grow, of course, any more than populations do. But they are structured to grow, and when they do, they grow exponentially.

There are many other factors in our society that may have the capacity for exponential growth. Violence may be inherently exponential, and corruption seems to feed on itself. Climate change also involves a variety of positive feedback. For example, emissions of greenhouse gases to the atmosphere lead to higher temperature, which in turn accelerates the melting of the Arctic tundra. As the tundra defrosts, it releases trapped methane. Methane is a potent greenhouse gas that can boost global temperatures even higher. Some positive feedbacks are explicitly included in

World3. We have modeled the forces that influence soil fertility. And various technologies appear to grow exponentially; we experiment with them in chapter 7. We believe, however, that growth processes governing population and industry have been the principal forces driving the global society past its limits, and we will focus on them.

Population and productive capital are the motors of exponential growth in human society. Other entities, such as food production, resource use, and pollution, tend to increase exponentially—not because they multiply themselves, but because they are *driven by* population and capital. There is no self-generation, no positive feedback loop, to cause pesticides in groundwater to create more pesticides, nor coal to breed underground and create more coal. The physical and biological consequences of growing 6 tons of wheat per hectare do not make it easier to grow 12 tons per hectare. At some point—when limits are reached—each doubling of food grown or minerals extracted is not easier but more difficult than the doubling before.

Therefore, insofar as food production and materials and energy use have been growing exponentially (which they have), they have been doing so not through their own structural capacity, but because the exponentially growing population and economy have been demanding more food and materials and energy and have been successful at producing them. Similarly, pollution and waste have been growing not because they have their own positive feedback structure, but because of the rising quantities of materials moved and energy used by the human economy.

A central assumption of the World3 model is that population and capital are structurally capable of exponential growth. This is not an arbitrary assumption. It is supported by the observable characteristics of the global socioeconomic system and by historical patterns of change. Growth in population and capital generates growth in the human ecological footprint unless or until there are profound changes in consumption preferences and drastic improvements in efficiencies of resource use. Neither change has yet occurred. The human population and capital plant and the energy and material flows that sustain them have grown exponentially for at least a century—though not smoothly, not simply, and not without strong impacts from other feedback loops. The world is more complicated than that. So is the World3 model, as we shall see.

World Population Growth

In the year 1650 the human population numbered around half a billion. It was growing at about 0.3 percent per year, with a doubling time of nearly 240 years.

By 1900 the population had reached 1.6 billion and was growing at 0.7 to 0.8 percent per year, a doubling time of about 100 years.

By 1965 the population totaled 3.3 billion. The rate of growth had increased to 2 percent per year, a doubling time of about 36 years. Thus the population grew not only exponentially from 1650, but in fact *super*exponentially—the rate of growth was itself growing. It was growing for a happy reason: Death rates were falling. Birth rates were also falling, but more slowly. Therefore the population surged.

After 1965 death rates continued to fall, but birth rates on average fell even faster (figure 2-4). While the population rose from 3.3 billion to just over 6 billion by the year 2000, the *rate* of growth fell from 2 to 1.2 percent per year.[6]

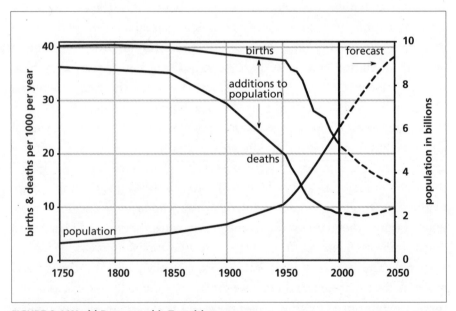

FIGURE 2-4 World Demographic Transition

The gap between births and deaths determines the rate at which population grows. Until about 1965 the average human death rate was dropping faster than the birth rate, so the population growth rate was increasing. Since 1965 the average birth rate has dropped faster than the death rate. Therefore the rate of population growth has decreased considerably—though the growth continues to be exponential. (Source: UN.)

That turnaround in population growth rate is an amazing shift, indicating major changes in the cultural factors that cause people to choose their family size and in the technical factors that enable them to carry out that choice effectively. The global average number of children born per woman went down from 5 in the 1950s to 2.7 in the 1990s. In Europe at the turn of the twenty-first century, completed family size averaged 1.4 children per couple, considerably less than the number required to replace the population.[6] The European population is projected to decline slowly, from 728 million in 1998 to 715 million in 2025.[7]

This fertility downturn does not mean that total world population growth has ceased, or ceased being exponential. It simply means that the doubling time has lengthened (from 36 years at 2 percent per year to 60 years at 1.2 percent per year) and may lengthen still farther. The net number of people added to the planet was in fact higher in 2000 than it was in 1965, though the growth rate was lower. Table 2-3 shows why: The lower rate in 2000 was multiplied by a larger population.

TABLE 2-3 Additions to World Population

Year	Population (million)	x	Growth Rate (% per year)	=	People Added (million per year)
1965	3,330	x	2.03	=	68
1970	3,690	x	1.93	=	71
1975	4,070	x	1.71	=	70
1980	4,430	x	1.70	=	75
1985	4,820	x	1.71	=	82
1990	5,250	x	1.49	=	78
1995	5,660	x	1.35	=	76
2000	6,060	x	1.23	=	75

(Source: UN.)

The annual number added to the world population finally did stop growing in the late 1980s. But the increase of 75 million in 2000 was still equivalent to adding in that year the total population of more than nine New York Cities every year. More accurately, since nearly all the increase took place in the South, it was equivalent to adding in one year the total population of the Philippines—or about 10 Beijings or six Calcuttas. Even with optimistic projections about further declines in the birth rate, a large population increase is still ahead, especially for the less industrialized countries (figure 2-5).

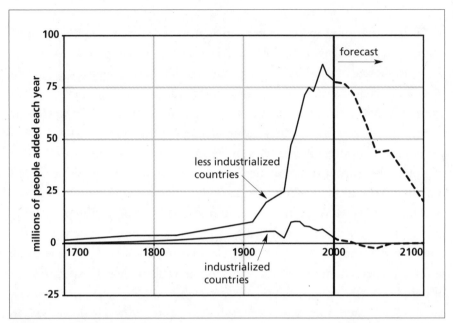

FIGURE 2-5 World Annual Population Increase
Until recently the number of people added to the world population each year had increased. Under the
UN's forecasts, that annual addition will soon drop steeply. Those forecasts assume rapid drops in birth
rates in the less industrialized countries. (Source: UN; D. Bogue)

The central feedback structure that governs the population system is
shown below.

On the left is the positive loop that can produce exponential growth. The
larger the population, the more births per year. On the right is a *negative feed-*

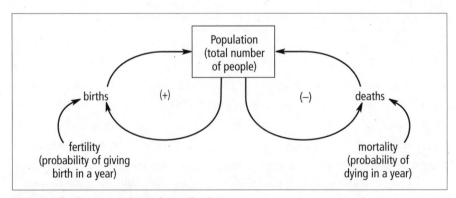

Birth Feedback Loop and Death Feedback Loop

back loop. Whereas positive loops generate runaway growth, negative loops tend to regulate growth, to hold a system within some acceptable range, or to return it to a stable state in which the system stocks have more or less constant values over time. A negative feedback loop propagates the consequences of a change in one element around the circle until they come back to change that element in a direction *opposite* to the initial change.

The number of deaths each year equals the total population times the average mortality—the average probability of death. The number of births equals the total population times the average fertility. The growth rate of a population is equal to its fertility minus its mortality. Of course, neither fertility nor mortality is constant. They depend upon economic, environmental, and demographic factors such as income, education, health care, family planning technologies, religion, pollution levels, and the population's age structure.

The most widespread theory about *how* fertility and mortality change and *why* global population growth rates are falling—the theory that is built into the World3 model—is called the *demographic transition*. According to this theory, in pre-industrial societies both fertility and mortality are high, and population growth is slow. As nutrition and health services improve, death rates fall. Birth rates lag by a generation or two, opening a gap between fertility and mortality that produces rapid population growth. Finally, as lives and lifestyles evolve into the patterns of a fully industrial society, birth rates fall, too, and the population growth rate slows.

The actual demographic experiences of 6 countries are shown in figure 2-6. You can see that birth and death rates in long-industrialized countries such as Sweden fell very slowly. The gap between them was never very large; the population never grew at more than 2 percent per year. Over the entire demographic transition, the populations of most countries of the North grew by a factor of five at most. By the year 2000 few industrial nations had a fertility level above replacement, and thus most are facing declining populations in the years to come. Those that were still growing were doing so because of immigration, demographic momentum (more young people coming into reproductive age than older people leaving it), or both.

In the South, where death rates fell later and faster, a large gap opened up between birth and death rates. This part of the world has experienced

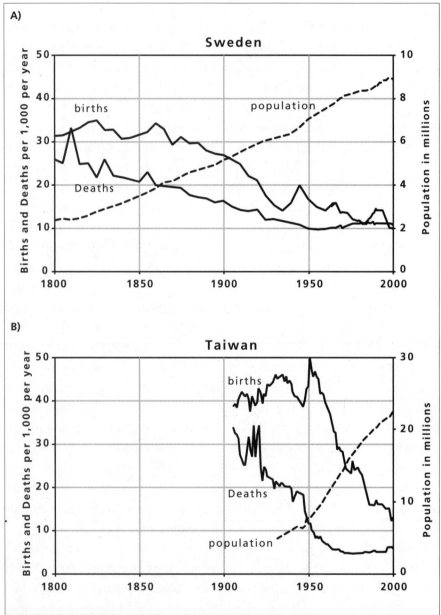

FIGURE 2-6 Demographic Transitions in Industrialized Countries (A) and in Less Industrialized Countries (B)

In the demographic transition a nation's death rate falls first, followed by its birth rate. Sweden's demographic transition occurred over almost 200 years, with the birth rate remaining rather close to the death rate. During this time Sweden's population increased less than fivefold. Japan is an example of a nation that has effected the transition in less than a century. The less industrialized countries of the late 1900s have experienced much larger gaps between their birth and death rates than any that ever prevailed in the now industrialized countries. (Sources: N. Keyfitz and W. Flieger; J. Chesnais; UN; PRB; UK ONS; Republic of China.)

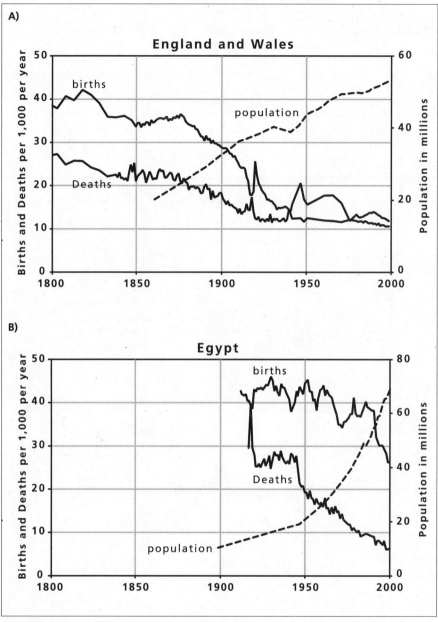

FIGURE 2-6 Demographic Transitions in Industrialized Countries (A) and in Less Industrialized Countries (B)

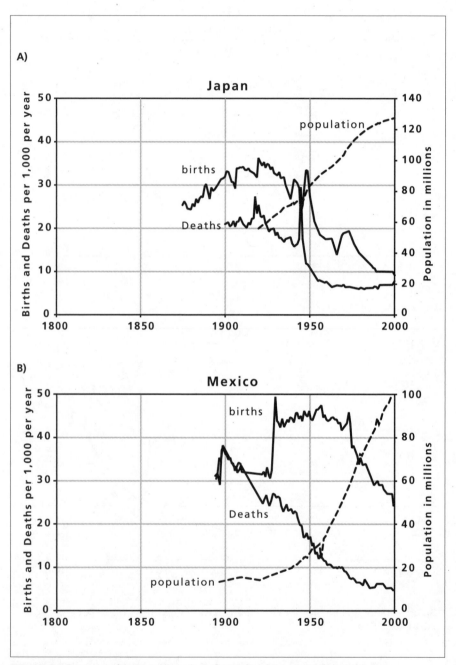

FIGURE 2-6 Demographic Transitions in Industrialized Countries (A) and in Less Industrialized Countries (B)

rates of population growth much greater than any the North ever had to deal with (except for North America, which absorbed high rates of immigration from Europe). The populations of many countries of the South have already grown by factors of 10 and are still growing. Their demographic transitions are far from complete.

Demographers argue about *why* there appears to be a demographic transition linked to industrialization. The driving factors are more complicated than rising income alone. Figure 2-7 shows, for example, the correlation between per capita income (measured as the gross national income, or GNI,[8] per person per year) and birth rates in various countries of the world. Clearly there is a strong relationship between high incomes and low birth rates. Just as clearly, especially at low incomes, there are striking exceptions. China, for example, has anomalously low birth rates for its level of income. Some Middle Eastern and African countries have anomalously high birth rates for theirs.

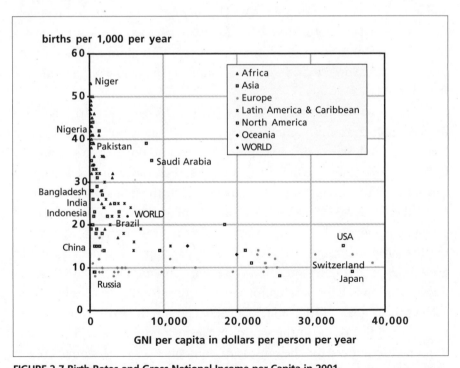

FIGURE 2-7 Birth Rates and Gross National Income per Capita in 2001
As a society becomes wealthier, the birth rate of its people tends to decline. The poorest nations experience birth rates from 20 to more than 50 births per 1,000 people per year. None of the richest nations has a birth rate above 20 per 1,000 per year. (Source: PRB; World Bank.)

The factors believed to be most *directly* important in lowering birth rates are not so much the size or wealth of the economy, but the extent to which economic improvement actually touches the lives of all families, and especially the lives of women. More important predictors than GNI per capita are factors such as education and employment (especially for women), family planning, low infant mortality, and relatively egalitarian distribution of income and opportunity.[9] China, Sri Lanka, Costa Rica, Singapore, Thailand, Malaysia, and several other countries have shown that when literacy, basic health care, and family planning are made available to most families, birth rates can drop even at modest income levels.

The World3 model contains many countervailing pressures on birth rates. We assume that a wealthier economy provides better nutrition and health care, which bring down death rates, and that it also improves family planning and reduces infant mortality, which bring down birth rates. We assume that industrialization reduces desired family size, over the long term and after a delay, by raising the cost of bringing up children and reducing their immediate economic benefits for their parents. We assume that a short-term income increase allows families to afford more children, within the range of children they desire, and a short-term income stagnation does the opposite.[10]

In other words, the model assumes and usually generates the long-term demographic transition, modulated by small short-term responses to increasing and decreasing income. The model population's tendency toward exponential growth is first enhanced, then moderated, by the pressures, opportunities, technologies, and norms of the industrial revolution.

In the "real world" at the turn of the millennium, population is still growing exponentially, though the rate of growth is dropping. The causes of that drop are more complicated than per capita income. Economic growth does not guarantee improvements in human welfare, greater freedom of choice for women, or lower birth rates. But it certainly helps achieve those goals. With some notable exceptions, the world's lowest birth rates do tend to occur in the world's richest economies. It is thus doubly important to understand the causes and consequences of economic growth in the World3 model and in the world.

World Industrial Growth

Public discussions of economic matters are full of confusion, much of which comes from a failure to distinguish between money and the real things money stands for.[11] We need to make those distinctions carefully here. Figure 2-8 shows how we represent the economy in World3, how we will talk about it in this book, and how, we believe, it is useful to think about the economy in a time of natural limits. Our emphasis is on the *physical economy*, the real things to which the earth's limits apply, not the *money economy*, which is a social invention not constrained by the physical laws of the planet.

Industrial capital refers here to actual hardware—the machines and factories that produce manufactured products. (With the help, of course, of

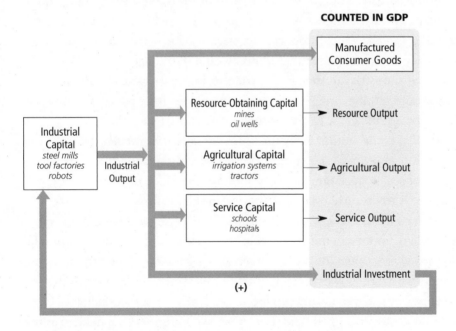

FIGURE 2-8 Flows of Physical Capital in the Economy of World3
The production and allocation of industrial output are central to the behavior of the simulated economy in World3. The amount of industrial capital determines how much industrial output can be produced each year. This output is allocated among five sectors in a way that depends on the goals and needs of the population. Some industrial capital is consumed; some goes to the resource sector to secure raw materials. Some goes to agriculture to develop land and raise land yield. Some is invested in social services, and the rest is invested in industry to offset depreciation and raise the industrial capital stock further.

labor, energy, materials, land, water, technology, finance, management, and the services of natural ecosystems and biogeochemical flows of the planet. We will come back to these co-factors of production in the next chapter.) We call the stream of real products (consumer goods and investment goods) made by industrial capital *industrial output*.

Some industrial output is in the form of equipment or buildings for hospitals, schools, banks, and retail stores. We call that *service capital*. Service capital produces its own stream of output that is nonmaterial but that has real value—such as health care and education.

Another type of industrial output is *agricultural capital*—tractors, barns, irrigation systems, harvesters—which produces *agricultural output*, mainly food and fiber.

Some industrial output takes the form of drills, oil wells, mining equipment, pipelines, pumps, tankers, refineries, and smelters. All that is *resource-obtaining capital*, which produces the stream of raw materials and energy necessary to allow all the other kinds of capital to function.

Some industrial output is classified as *consumer goods*—clothing, cars, radios, refrigerators, houses. The amount of consumer goods per person is an important indicator of the population's material well-being.

Finally, some output is in the form of *industrial capital*. This we call *investment*—steel mills, electric generators, lathes, and other machines, which offset depreciation and may increase the stock of industrial capital, enabling it to produce even more output in the future.

So far everything we have mentioned here is physical stuff, not money. The role of money in the "real world" is to convey information about relative costs and values of stuff (values as assigned by producers and consumers who have power in the market). Money mediates and motivates the flows of physical capital and products. The annual money value of all physical outputs of final goods and services shown in figure 2-8 is defined as the GDP, the gross domestic product.

We will refer to GDP in various figures and tables, because the world's economic data are mainly expressed in money terms, not physical terms. But our interest is in what GDP *stands for*: the real capital stocks, industrial goods, services, resources, agricultural products, and consumer goods. This stuff, not the dollars, allows the economy and society to function. This stuff, not the dollars, is extracted from the planet, and eventually it goes back to the planet through disposal into the soil, air, or water.

We have already said that industrial capital can grow exponentially by its own self-reproduction. The feedback structure representing that self-generation is similar to the one we drew for the population system.

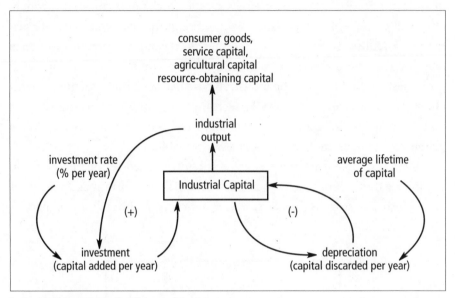

Industrial Capital Feedback Loop Structure

A given amount of industrial capital (factories, trucks, computers, power plants) can produce a certain amount of manufactured output each year, as long as other necessary inputs are sufficient. Some percent of each year's production is investment—looms, motors, conveyer belts, steel, cement—which goes to increase the capital stock and thereby expand the capacity for production in the future. This is the "birth rate" of capital. The fraction invested is variable, just as human fertility is variable, depending on decisions, desires, and constraints. There are delays in this positive feedback loop, since the planning, financing, and construction time for a major piece of capital equipment such as a railroad, electric generating plant, or refinery can take years or even decades.

Capital, like population, has a "death loop" as well as a "birth loop." As machines and factories wear out or become technically obsolete, they are shut down, dismantled, recycled, discarded. The rate of capital depreciation is analogous to the death rate in the population system. The more capital is present, the more there is to wear out each year, so the less there will be the next year, unless the inflow of new investment is sufficient to replace depreciated capital.

Just as populations undergo a demographic transition during the process of industrialization, so an economy's capital stocks pass through a widely observed pattern of growth and change. Pre-industrial economies are primarily agricultural and service economies. As the capital growth loop starts operating, all economic sectors grow, but for a while the industrial sector grows fastest. Later, when the industrial base has been built, further growth takes place primarily in the service sector (see figure 2-9). This transition is built into the World3 model as its default mode of economic growth, unless deliberate changes are made to test other possibilities.[12]

Highly developed economies are sometimes called service economies, but in fact they continue to require a substantial agricultural and industrial base. Hospitals, schools, banks, stores, restaurants, and hotels are all part of

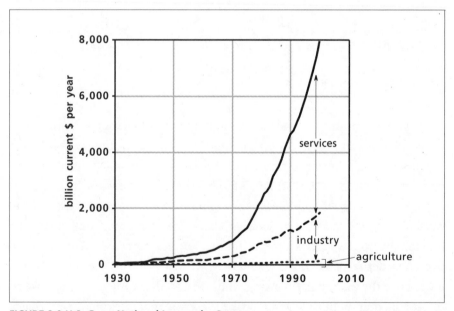

FIGURE 2-9 U.S. Gross National Income by Sector
The history of the distribution of the value of the U.S. economic output among service, industry, and agriculture shows the transition to a service economy. Note that although services assume the largest share of the economy, the industrial and agricultural sectors continue to expand in absolute terms. (Source: U.S. Bureau of Economic Analysis.)

the service sector. Watch delivery trucks bringing them food, paper, fuel, and equipment, or garbage trucks hauling their waste away. Measure what goes down their drainpipes and up their chimneys, and you will know that service sector enterprises need a constant, hefty stream of physical throughput from the earth's sources to the earth's sinks. Along with industries, they make a significant contribution to the ecological footprint of humanity.

Steel mills and mines may be located far away from the offices of the information economy. The tonnage of materials used may not rise as fast as the dollar value of output. But as figure 2-9 shows, even in a "post-industrial" economy, the industrial base does not decline. Information is a wonderful, valuable, disembodied commodity, but it is typically stored in a desktop computer that, as of 1997, was made from 55 pounds of plastic, metal, glass, and silicon; that drew 150 watts of electricity; and that generated in its manufacture 139 pounds of waste materials.[13] The people who produce, process, and use information not only eat food but also drive cars, live in houses, work in heated or cooled buildings, and—even in the age of electronic communications—use and discard reams of paper.

The positive loop causing growth in the world capital system has operated to make industry grow faster than the population. From 1930 to 2000 the money value of world industrial output grew by a factor of 14 (as shown in figure 1-2). If the population had been constant over that period, the material standard of living would have grown by a factor of 14 as well, but because of population growth the average per capita output increased by a factor of 5. Between 1975 and 2000 the size of the industrial economy roughly doubled, while output per capita rose only about 30 percent.

More People, More Poverty, More People

Growth is necessary to end poverty. That seems obvious. Less obvious to its many proponents is the fact that growth in the economic system, as it is currently structured, will not end poverty. On the contrary, current modes of growth perpetuate poverty and increase the gap between the rich and the poor. In 1998 more than 45 percent of the globe's people had to live on incomes averaging $2 a day or less. That is more poor people than there were in 1990, even after a decade that saw astonishing income gains for many.[14] The fourteenfold increase in world industrial output since 1930 has made some people very wealthy, but it has not ended poverty. There is no reason to expect that another fourteenfold increase (if it were possible within the earthly limits) would end poverty, unless the global system were restructured to direct growth to those who most need it.

In the current system economic growth generally takes place in the already rich countries and flows disproportionately to the richest people within those countries. Figure 2-10 shows GNI per capita growth curves for the world's 10 largest nations (by population) plus the European Union. They illustrate how decades of growth have systematically increased the gap between rich countries and the poor ones.

According to the United Nations Development Program, in 1960 the 20 percent of the world's people who lived in the wealthiest nations had 30 times the per capita income of the 20 percent who lived in the poorest

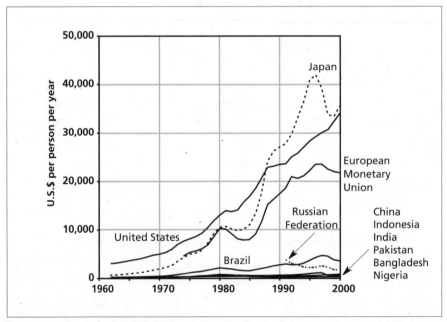

FIGURE 2-10 Per Capita GNI of the Top 10 Most Populous Countries and the European Monetary Union

Economic growth takes place primarily in the nations that are already rich. The 6 countries of Indonesia, China, Pakistan, India, Bangladesh, and Nigeria together contain almost half the world's population. Their per capita GNI barely rises off the axis when plotted together with the GNI per capita of the wealthier nations. (Source: World Bank.)

nations. By 1995 the average income ratio between the richest and the poorest 20 percent had increased from 30:1 to 82:1. In Brazil the poorest half of the population received 18 percent of the national income in 1960 and only 12 percent in 1995. The richest 10 percent of Brazilians received 54 percent of national income in 1960, rising to 63 percent in 1995.[15] The average African household consumed 20 percent less in 1997 than it did in 1972.[16] A century of economic growth has left the world with enormous disparities between the rich and the poor. One indicator of this, share of gross world product by different income groups, is shown in figure 2-11.

When we, system dynamicists, see a pattern persist in many parts of a system over long periods, we assume that it has causes embedded in the feedback loop structure of the system. Running the same system harder or faster will not change the pattern as long as the structure is not revised. Growth as usual has widened the gap between the rich and the poor.

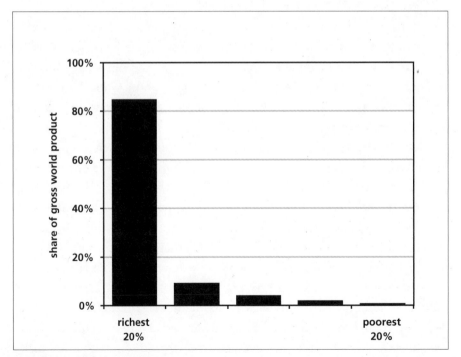

FIGURE 2-11 Global Disparities

The global distribution of wealth and opportunities is extremely skewed. The richest 20 percent of the world's population controls more than 80 percent of the world gross product and uses nearly 60 percent of world commercial energy. (Source: World Bank.)

Continuing growth as usual will never close that gap. Only changing the structure of the system—the chains of causes and effects—will do that.

What is the structure that keeps widening the gap between the rich and the poor even in the presence of enormous economic growth? We see two generic structures at work. The first has to do with social arrangements— some common in many cultures, some unique to particular cultures—that *systemically reward the privileged with the power and resources to acquire even more privilege.* Examples range from overt or covert ethnic discrimination to tax loopholes for the wealthy; from inferior nutrition for the children of the poor to premium schooling for the children of the wealthy; from the use of money to gain political access, even in supposed democracies, to the simple fact that interest payments systematically flow from those who have less money than they need to those who have more than they need.

In systems terms these structures are called "success to the successful" feedback loops.[17] They are positive loops that reward the successful with the means to succeed. They tend to be endemic in any society that does not consciously implement counterbalancing structures to level the playing field. (Examples of counterbalancing structures include anti-discrimination laws, tax rates that increase as a person grows richer, universal education and health care standards, "safety nets" to support those who fall upon hard times, taxes on wealth, and democratic processes that separate politics from the influence of money.)

None of these "success to the successful" loops is explicitly represented in the World3 model. World3 is not a model of the dynamics of income or wealth or power distribution; its focus is on the aggregate relationship between the world economy and the limits to growth.[18] So it assumes a continuation of current distribution patterns.

There is, however, one structure in World3 that reflects the links between the population and capital systems as we have described them in this chapter. This structure perpetuates poverty, population growth, and the tendency of the world system to overshoot its limits. It must be changed, as we will demonstrate in later chapters, if a sustainable world is to be achieved.

This poverty-perpetuating structure arises from the fact that it's easier for rich populations to save, invest, and multiply their capital than it is for poor ones to do so. Not only do the rich have greater power to control market conditions, purchase new technologies, and command resources, but centuries of growth have built up for them a large stock of capital that multiplies itself. Most basic needs are met, so relatively high investment rates are possible without depriving the present population of essentials. Low population growth permits more output to be allocated to achieving economic growth and less to meeting the health and education needs of a rapidly expanding populace.

In poor countries, by contrast, capital growth has a hard time keeping up with population growth. Output that might be reinvested is more likely to be required to provide schools and hospitals and to fulfill subsistence consumption needs. Because immediate requirements leaves little output for industrial investment, the economy grows only slowly. The demographic transition stays stuck in the middle phase, with a large gap between birth and death rates. When women see no attractive educational or economic alternatives to childbearing, children are one of the few forms of investment available; thus the population grows bigger without growing richer. As the saying goes, "The rich get richer and the poor get children."

International gatherings can become paralyzed by passionate arguments about which arrow in this feedback loop is most important: Poverty causes population growth, or population growth causes poverty.

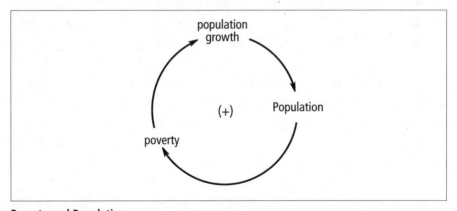

Poverty and Population

In fact, all parts of this positive feedback loop have a strong influence on the behavior of populations in poorer areas. They form a "system trap," a "less success to the already unsuccessful" loop, that keeps the poor poor and the population growing. By drawing output away from investment and into consumption, population growth slows capital growth. Poverty, in turn, perpetuates population growth by keeping people in conditions where they have no education, no health care, no family planning, no choices, no power, no way to get ahead except to hope their children can bring in income or help with family labor.

One consequence of this trap is shown in figure 2-12. Food production in every part of the South has increased greatly over the past 20 years. In most places it has doubled or tripled. But because of rapid population growth, food production per person has barely improved, and in Africa it has steadily decreased. The only places where food production has noticeably kept ahead of population growth are Europe and the Far East.

The graphs in figure 2-12 show a double tragedy. The first is a human one. A great agricultural achievement, a tremendous increase in food production, has been largely absorbed not in feeding people more adequately but in feeding more people inadequately. The second tragedy is environmental. The increase in food production has been attained by policies that damaged soils, waters, forests, and ecosystems, a cost that will make future production increases more difficult.

But any positive feedback loop that grinds a system down can be turned around to lift the system up. More poverty means more population, which means more poverty. But less poverty means slower population growth, which means less poverty. With enough investment sustained for a long enough time, with fair pricing for products and labor, with increasing output allocated much more directly to the poor, and especially to the education and employment of women and to family planning, the effects of the population–poverty loop can be reversed. Social improvements can reduce the population growth rate. That can allow more investment in industrial capital, which produces more goods and services. Rising consumption of goods and services helps reduce population growth still farther.

In parts of the world where there is careful attention to the welfare of the whole population, and especially the poor, that turnaround is happening. This is one reason why the world population growth rate is falling, and the demographic transition is proceeding.

But in other places, where inequity is culturally endemic, where the

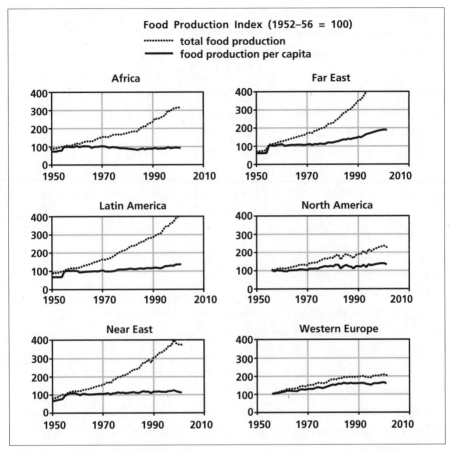

FIGURE 2-12 Regional Food Production
The index of total food production (index = 100 in 1952–56) has doubled or tripled in the past 50 years in the regions of the world where hunger is greatest, but the index of food production per person has scarcely changed in those areas because population has grown almost as fast. In the case of Africa, per capita food production declined by 9 percent between 1996 and 2001. (Source: FAO.)

resources or the will to invest in the public welfare are lacking, or where monetary failures have brought the imposition of "structural adjustments" that divert investment from education and health care systems, there is no widespread improvement in the lives of the people. Stuck in poverty and still growing rapidly, those populations are in grave danger of having their growth cut off not by falling birth rates but by rising death rates. Indeed, Zimbabwe, Botswana, Namibia, Zambia, and Swaziland are expected to reach zero population growth early in the twenty-first century for a tragic reason—the deaths of young adults and children from AIDS.[19]

Exponential growth of population and industrial production is built into the self-generating structure of the "real world" socioeconomic system, but in a complicated way that tends to swing some parts of the world toward slow population growth and fast industrial growth, and other parts toward slow industrial growth and fast population growth. But in both cases population and physical capital keep growing.

Can this physical growth realistically continue forever? Our answer is no! Growth in population and capital increases the ecological footprint of humanity, the burden humanity places on the world ecosystem, unless there is a successful effort to avoid such an increase. In principle, it is possible to reduce the ecological footprint of each unit of human activity (through technological and other means) quickly enough to allow continuing growth in population and industrial capital. But we do not believe this will be achieved in practice. Certainly the empirical evidence available from around the world today shows that a sufficient reduction is not taking place. The ecological footprint is still growing (see figure P-1 in the authors' preface), albeit at a slower pace than the economy.

Once the footprint has grown beyond the sustainable level, as it already has, it must eventually come down—either through a managed process (for example, through rapid increases in eco-efficiency) or through the work of nature (say, through declining use of wood as forests disappear). There is no question about whether growth in the ecological footprint will stop; the only questions are when and by what means.

Population growth will essentially cease, either because the birth rates fall farther, or because deaths begin to rise—or both. Industrial growth will essentially cease, either because investment rates fall, or depreciation begins to rise—or both. If we anticipate these trends, we may exert some rational control over them, selecting the best of the options available to us. If we ignore them, then the natural systems will choose an outcome without regards to human welfare.

Birth and death rates, investment and depreciation rates will be balanced by human choice or by feedback from overstressed earthly sources and sinks. The exponential growth curves will slow, bend, and either level off or decline. The condition of human society and the planet at that point could be disastrous.

It's entirely too easy to classify things as "bad" or "good" and to keep those classifications fixed. For generations both population growth and capital growth were classified as an unmitigated good. On a lightly populated planet with abundant resources, there were good reasons for that positive valuation. Now, with an ever clearer understanding of ecological limits, it can be tempting to classify all growth as bad.

The task of managing in an era of limits demands greater subtlety, more careful classification. Some people desperately need more food, shelter, and material goods. Some people, in a different kind of desperation, try to use material growth to satisfy other needs, which are also very real but nonmaterial—needs for acceptance, self-importance, community, identity. It makes no sense, therefore, to talk about growth with either unquestioning approval or unquestioning disapproval. Instead it is necessary to ask: *Growth of what? For whom? At what cost? Paid by whom? What is the real need here, and what is the most direct and efficient way for those who have that need to satisfy it? How much is enough? What are the obligations to share?*

The answers to those questions can point the way toward a sufficient and equitable society. Other questions will point the way toward a sustainable society. *How many people can be provided for with a given throughput stream—within a given ecological footprint? At what level of material consumption? For how long? How stressed is the physical system that supports the human population, the economy, and all other species? How resilient is that support system to what kinds and quantities of stress? How much is too much?*

To answer those questions, we must turn our attention from the causes of growth and focus on the limits to growth. That is our goal for chapter 3.

CHAPTER 3

The Limits: Sources and Sinks

The technologies we adopted that enabled us to maintain constant or declining dollar costs for resources often required ever-increasing amounts of direct and indirect fuel . . . this luxury becomes a costly necessity, requiring that increasing proportions of our national income be diverted to the resource-processing sectors in order to supply the same quantity of resource.

—World Commission on Environment and Development, 1987

Our concern about collapse does not come from belief that the world is about to exhaust the planet's stocks of energy and raw materials. Every scenario produced by World3 shows that the world in the year 2100 still has a significant fraction of the resources that it had in the year 1900. In analyzing World3 projections our concern rather arises from the growing cost of exploiting the globe's sources and sinks. Data on these costs are inadequate, and there is substantial debate on the issue. But we conclude from the evidence that growth in the harvest of renewable resources, depletion of nonrenewable materials, and filling of the sinks are combining slowly and inexorably to raise the amount of energy and capital required to sustain the quantity and the quality of material flows required by the economy. Those costs arise from a combination of physical, environmental, and social factors. Eventually they will be high enough that growth in industry can no longer be sustained. When that happens, the positive feedback loop that produced expansion in the material economy will reverse direction; the economy will begin to contract.

We cannot prove this assertion. We can try to make it plausible, and then point out the nature of constructive responses. To accomplish that in this chapter we present a large amount of information about sources and sinks. We summarize the situation and the prospects for a variety of resources

51

that will be needed to sustain world economic and population growth for the coming century. The list of necessary inputs is long and diverse, but it can be divided into two main categories.

The first category includes the physical necessities that support all biological and industrial activity—fertile land, minerals, metals, energy, and the ecological systems of the planet that absorb wastes moderate climate. These ingredients are, in principle, tangible, countable items, such as hectares of arable land and forests, cubic kilometers of fresh water, tons of metals, billions of barrels of petroleum. These ingredients are, in practice, however, remarkably difficult to quantify. Their total quantities are uncertain. They interact—some can substitute for others while producing some makes it more difficult to obtain others. The definitions of *resources, reserves, consumption,* and *production* are inconsistent; the science is incomplete, and bureaucracies often distort or hide the numbers for their own political and economic ends. And information about physical realities is typically expressed with economic indices, such as monetary price. Prices are determined in markets, and they observe a set of rules very, very different from those that govern physical resources. Nevertheless, we will focus on these physical necessities in this chapter.

The second category of requirements for growth consists of the social necessities. Even if the earth's physical systems are capable of supporting a much larger, more industrially developed population, the actual growth of the economy and the population will depend on such factors as peace and social stability, equity and personal security, honest and far sighted leaders, education and openness to new ideas, willingness to admit mistakes and to experiment, and the institutional foundations for steady and appropriate technical progress.

These social factors are difficult to assess and probably impossible to predict with useful accuracy. Neither this book nor World3 deals explicitly with these social factors in a detailed, useful way. We lack the data and the causal theories required to incorporate them into our formal analysis. But we know that while fertile land, sufficient energy, necessary resources, and a healthy environment are necessary for growth, they are not sufficient. Even if they are physically abundant, their availability may be curtailed by social problems. We assume here, however, that the best possible social conditions will prevail.

The materials and energy used by the population and capital plant do not come from nowhere. They are extracted from the planet. And they do not disappear. When their economic use is over, materials are recycled or become wastes and pollutants; energy is dissipated as unuseable heat. Streams of material and energy flow from the planetary *sources* through the *economic subsystem* to the planetary *sinks* where wastes and pollutants end up (figure 3-1). Recycling and cleaner production can dramatically reduce, but never eliminate, the waste and pollution per unit of consumption. People will always need some food, water, clean air, shelter, and many kinds of materials to grow, to maintain healthy bodies, to live productive lives, and to generate both capital and new people. Machines and buildings will always need some energy, water, air, and a variety of metals, chemicals, and biological materials to produce goods and services, to be repaired, and to make more machines and buildings. There are limits to the rates at which the sources can produce

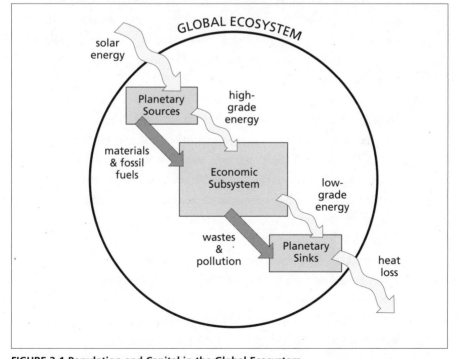

FIGURE 3-1 Population and Capital in the Global Ecosystem
Population and capital are sustained by flows of fuels and nonrenewable resources from the planet, and they produce outflows of heat and waste, which contaminate the air, water, and soil of the planet. (Source: R. Goodland, H. Daly, and S. El Serafy.)

and the sinks absorb these flows without harm to people, the economy, or the earth's processes of regeneration and regulation.

The nature of these limits is complex, because the sources and sinks are themselves part of a dynamic, interlinked system, maintained by the bio-geochemical cycles of the planet. There are short-term limits (the amount of oil refined and waiting in storage tanks, for instance) and long-term limits (the amount of accessible oil under the ground). Sources and sinks may interact, and the same natural system may serve as both source and sink. A plot of soil, for example, may be a source for food crops and a sink for acid rain caused by air pollution. Its capacity to serve either of those functions may depend on the extent to which it is serving the other.

Economist Herman Daly has suggested three simple rules to help define the sustainable limits to material and energy throughput:[1]

- For a *renewable resource*—soil, water, forest, fish—the sustainable rate of use can be no greater than the rate of regeneration of its source. (Thus, for example, fish are harvested unsustainably when they are caught at a rate greater than the rate of growth of the remaining fish population.)

- For a *nonrenewable resource*—fossil fuel, high-grade mineral ores, fossil groundwater—the sustainable rate of use can be no greater than the rate at which a renewable resource, used sustainably, can be substituted for it. (For example, an oil deposit would be used sustainably if part of the profits from it were systematically invested in wind farms, photovoltaic arrays, and tree planting, so that when the oil is gone, an equivalent stream of renewable energy is still available.)

- For a *pollutant* the sustainable rate of emission can be no greater than the rate at which that pollutant can be recycled, absorbed, or rendered harmless in its sink. (For example, sewage can be put into a stream or lake or underground aquifer sustainably no faster than bacteria and other organisms can absorb its nutrients without themselves overwhelming and destabilizing the aquatic ecosystem.)

Any activity that causes a renewable resource stock to fall, or a pollution sink to rise, or a nonrenewable resource stock to fall without a renewable replacement in sight, cannot be sustained. Sooner or later that activity will have to be diminished. In many discussions of the Daly rules—in academic, business, government, and civic circles—we have never heard anyone challenge them. (We've also rarely found anyone seriously trying to live by them.) If there are basic laws of sustainability, these must be among them. The question is not whether they are correct; the questions are whether the global economy is respecting them, and what happens if it doesn't.

We will use the three Daly criteria to make a quick survey of various sources and sinks used by the human economy. Starting with renewable resources, we ask: *Are they being used faster than they regenerate?* For nonrenewable resources, whose stocks by definition must be falling, we ask: *How quickly are the high quality materials being used. What is the course of the true costs in energy and capital required to provide them?* Finally, we turn to pollutants and wastes and ask: *Are they being rendered harmless at sufficient rates? Or are they accumulating in the environment?*

These questions are to be answered not with the World3 model (nothing in this chapter depends upon that model), but with the global data, insofar as those data exist, source by source, sink by sink.[2] In this chapter we will mention only a few of the many interactions of one source or sink with another (for example, the fact that growing more food takes more energy, or that pollution from producing more energy can change the climate and affect agricultural yields).

The limits we discuss here are among the ones the world's scientists presently know about. There is no guarantee that they are in fact the most important. There will be surprises ahead, pleasant and unpleasant. The technologies we mention here will certainly be improved in the future. On the other hand, new problems will become apparent that are totally unrecognized today.

We go into some detail about the status and prospects for the globe's physical necessities. Our analysis will not give you a simple, unambiguous view of where humanity is vis-à-vis the limits to growth. But it will help you form your own understanding about the reality of limits and the impact of current policies on them. Even after we make allowances for the present gaps in human understanding about limits, we think the evidence presented in this chapter will persuade you of four conclusions:

- The human economy is now using many critical resources and producing wastes at rates that are not sustainable. Sources are being depleted. Sinks are filling and, in some cases, overflowing. Most throughput streams cannot be maintained over the long term even at their current flow rates, much less increased. We expect many of them will reach their peaks and then decline in this century.
- These high rates of throughput are not necessary. Technical, distributional, and institutional changes could reduce them greatly while sustaining and even improving the average quality of life of the world's people.
- The human burden on the natural environment is already above sustainable levels, and it cannot be maintained for more than a generation or two. As a consequence, there are already apparent many negative impacts on human health and the economy.
- The true costs of materials are increasing.

The concept of human burden on the environment is exceedingly complex and difficult to quantify. The best current approach, the one we use here, is the notion of an ecological footprint. That concept is defined as the total impact of humanity on nature: the sum of all effects of resource extraction, pollution emission, energy use, biodiversity destruction, urbanization, and the other consequences of physical growth. It is a difficult concept to measure, but great advances have been made during the past decade. And they will certainly continue.

One promising approach, mentioned in the preface, is to convert all human draws on the global ecosystem into the equivalent number of hectares of global land necessary to sustain the "ecological service provided" indefinitely. There is a finite number of hectares of land on earth. Thus this approach provides one answer to the question of whether humanity exceeds the available resource supply. Figure P-1 in the authors' preface indicates that the answer is yes. According to this method of measuring the ecological footprint, at the turn of the millennium humanity required an area 1.2 times larger than the amount of land available on planet Earth. In short, humanity was 20 percent above the global limit. Fortunately, there are many ways to relieve the strain, to get back down below the limits and support human needs and hopes much more sustainably. We will discuss many of these ways in the following pages.[3]

Renewable Sources

Food, Land, Soil

Most high-quality agricultural land is already in production, and the environmental costs of converting remaining forest, grassland, and wetland habitats to cropland are well recognized. . . . Much of the remaining soil is less productive and more fragile. . . . One analysis of global soil erosion estimates that, depending on the region, topsoil is currently being lost 16 to 300 times faster than it can be replaced.

—WORLD RESOURCES INSTITUTE, 1998

Between 1950 and 2000 world grain production more than tripled, from around 590 to more than 2,000 million metric tons per year. From 1950 to 1975 grain output increased by an average of 3.3 percent per year, faster than the 1.9 percent per year rate of population growth (figure 3-2). Yet during the past few decades the rate of grain production increase has slowed until it has fallen below the population growth rate. Per capita grain production peaked in about 1985, and it has been falling slowly ever since.[4]

Still, there is enough food, at least in theory, to feed everyone adequately. The total amount of grain produced in the world around the year 2000 could keep eight billion people alive at subsistence level, if it were evenly distributed, not fed to animals, and not lost to pests or allowed to rot between harvest and consumption. Grain constitutes roughly half the world's agricultural output (measured in calories). Add the annual output of tubers, vegetables, fruits, fish, and animal products raised from grazing rather than grain, and there would be enough to give the turn-of-the-millennium population of six billion a varied and healthful diet.[5]

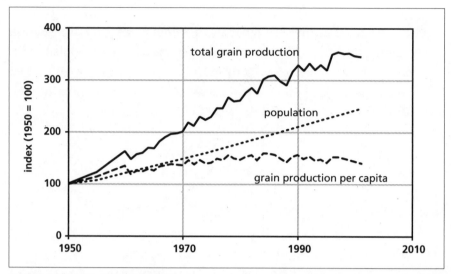

FIGURE 3-2 World Grain Production

The world's farmers produced more than three times as much grain in 2000 as in 1950. Because of population growth, however, per capita production rose to a peak in the mid-1980s and has been declining slightly since. Still, world grain production per capita is now 40 percent higher than it was in 1950. (Sources: FAO; PRB.)

Actual postharvest loss varies by crop and place, ranging from 10 to 40 percent.[6] The distribution of food is far from equal, and much grain goes to feed animals, not people. Thus, in the midst of theoretical adequacy, hunger persists. The United Nations Food and Agriculture Organization (FAO) estimates that about 850 million people chronically eat less food than their bodies require.[7]

These hungry people are primarily women and children. In the developing countries one of every three children is malnourished.[8] Some 200 million people in India are chronically hungry; more than 200 million in Africa; 40 million in Bangladesh; 15 million in Afghanistan.[9] Roughly nine million people die every year of causes related to hunger. That comes to an average of 25,000 deaths a day.

So far the number of hungry people has remained roughly constant as the population has grown. The estimated annual number of deaths from hunger has slowly dropped. That is an amazing accomplishment. In a world of growing population and pressing limits, hunger is not worsening. But there are still pockets of desperate hunger and more widespread areas of chronic malnourishment.

Hunger does not persist because of the earth's physical limits—not yet, anyway. More food could be raised. For example, figure 3-3 shows trends in grain yields in several countries and the world. Because of differences in soils and climate, every hectare of land cannot be expected to produce as much as the highest known yields in the most favorable places. But output could certainly increase in many places with techniques already well known and widely practiced.

In a thorough study of soils and climate in 117 countries of Latin America, Africa, and Asia, the FAO estimated that only 19 of these countries would *not* be able to feed their year-2000 populations from their own lands, if they could use every potentially arable hectare and get the highest yields technically possible. According to this study, if all cultivatable land were allocated to food, if there were no loss to erosion, if there were perfect weather, perfect management, and uninhibited use of agricultural inputs, the 117 countries studied could multiply their food output by a factor of 16.[10]

Of course those assumptions are wildly unrealistic. Given actual weather and farming practices, given the need to use land for purposes other than food production (purposes such as forests, pasture, human settlements, protection of watersheds, protection of biodiversity), given problems with fertilizer and pesticide runoff, the practical limits to food production are considerably lower than the theoretical limits. Indeed, as we have seen, per capita grain production has been falling since 1985.

> The period since World War II has seen remarkable growth in agricultural production and productivity in the developing world. While in many farming areas this growth has apparently been sustainable, in others it derived from two unsustainable processes: the clearing of new lands of lower productive potential or higher vulnerability, and the intensification of production by mining or destroying the soil resource base.[11]

The most obvious limit is land.[12] Estimates of potential cultivatable land on Earth range from two to four billion hectares (5 to 10 billion acres), depending on what is considered *cultivatable*. Roughly 1.5 billion hectares are actually cropped, an area that has been nearly constant for three

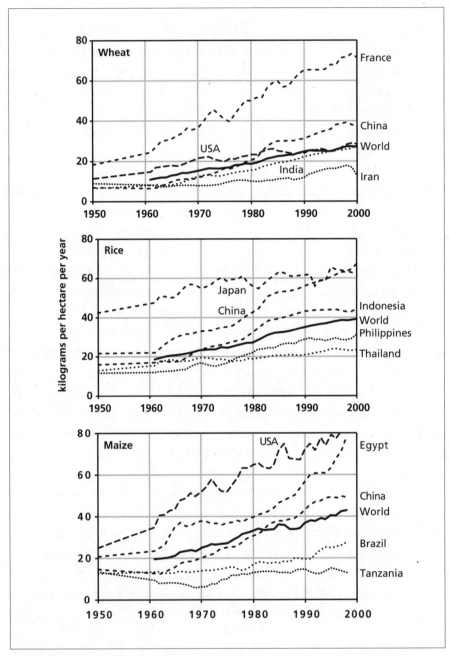

FIGURE 3-3 Grain Yields

Yields of wheat, rice, and maize (corn) are high in the industrialized world. In some industrializing nations, such as China, Egypt, and Indonesia, they are rising fast. In other less industrialized nations, yields are still very low, with considerable potential for improvement. (In order to smooth out yearly weather variations, yields in these graphs have been averaged over three-year intervals.) (Source: FAO.)

decades. Food production increases have come almost entirely from yield increases, not from a net expansion of land. But that does not mean the area of cultivated land is being sustained. New farmland is continually being brought into production, while once productive land is being lost to erosion, salt buildup, urbanization, and desertification. So far the losses have roughly balanced the additions, in area if not in quality. Since the best land is generally developed first, formerly prime soils are being degraded, while more marginal land is being brought into production.[13]

The United Nations Environment Program estimated in 1986 that over the past 1,000 years humans have turned about two billion hectares of productive farmland into wasteland.[14] That is more than the total cropped today. About 100 million hectares of irrigated land has been lost to salinization, with another 110 million experiencing reduced productivity. The rate of humus loss is accelerating, from 25 million tons per year prior to the industrial revolution, to 300 million tons per year over the past several centuries, to 760 million tons per year during the past 50 years.[15] This humus loss not only erodes soil fertility but also adds to the buildup of carbon dioxide in the atmosphere.

The first global assessment of soil loss based on comparable studies by several hundred regional experts was published in 1994. It concluded that 38 percent (562 million hectares) of currently used agricultural land has been degraded (plus 21 percent of permanent pasture and 18 percent of forestland).[16] The degree of degradation ranges from slight to severe.

We have not been able to find global figures for conversion of farmland to roads and settlements, but the loss must be substantial. The city of Jakarta is estimated to be expanding into cropland at a rate of 20,000 hectares per year. Vietnam has been losing 20,000 hectares per year of rice paddies to urban development. Thailand turned 34,000 hectares of agricultural land into golf courses between 1989 and 1994. China lost 6.5 million hectares of arable land to development between 1987 and 1992, but at the same time brought 3.8 million hectares of forest and pasture into cultivation. The United States paves over roughly 170,000 hectares of cropland per year.[17]

Because of such developments, two renewable sources are being drawn down. The first is the quality (depth, humus content, fertility) of soils on cultivated land. For a long time this loss may not be visible in food output, because soil nutrients can be replaced by nutrients from chemical fertilizers.[18] Fertilizers will mask the signs of soil abuse, but not indefinitely. They are themselves an unsustainable input to the agriculture system, introducing a delay in signals about soil fertility, one of the structural characteristics that leads to overshoot.

The second unsustainably used source is the land itself. If millions of hectares are being degraded and abandoned while the cultivated area remains roughly constant, that must mean the area of potentially arable land (mostly forest, as we shall see later in this chapter) is shrinking, while the area of unproductive wasteland is growing. The stream of food that sustains the human population is being produced by constantly moving onto new land while leaving behind exhausted, salted, eroded, or paved soils. Obviously that practice cannot go on forever.

If the population is growing exponentially and the area of cultivated land has been roughly constant, then the cultivated area per person is declining. In fact, it fell from 0.6 hectare per person in 1950 to 0.25 hectare per person in 2000. It has been possible to go on feeding a growing population with less land per person only because yields have been rising. The average hectare of rice yielded 2 tons per year in 1960 and 3.6 tons in 1995, with top yields, under experiment-station conditions, of 10 tons. Corn (maize) yields in the United States went up from an average of 5 tons per hectare in 1967 to more than 8 in 1997, with the top farmers in the best years achieving 20.

What do all these data suggest about the potential for future agricultural land scarcity? Figure 3-4 illustrates several land scenarios for the coming century. It shows the interrelation of total cultivated land, population growth, average yield, and standard of diet.

The shaded band indicates the total amount of cultivatable land ranging from the present 1.5 billion hectares to the theoretical upper limit of 4 billion hectares. Land at the top of the shaded band will be much less productive than that at the bottom. Of course total cultivated land could decline, but we will assume in figure 3-4 that no more land is lost. In each scenario we further assume that global population will grow according to the median forecast of the United Nations.

Clearly yield gains are coming more slowly and more expensively. Some American agriculture experts worried already in 1999 about a "yield plateau."[19] Erosion, climate change, costly fossil fuels, falling water tables,

> **It's a striking pattern. Steady progress upward on the average, but at the top—the best of the best—it doesn't appear that maize yields have changed in 25 years. Average annual maize yields keep right on going up by 90 kg/ha, but the investment in maize-breeding research has gone up four-fold. When every step forward is harder to take, that's a sign of diminishing returns.**
>
> Kenneth S. Cassman, 1999
>
> **I can't tell myself a convincing story about where the growth [in yield] is going to come from in the next half century.**
>
> Vernon Ruttan, 1999
>
> **Those maximum rice yields have been the same for 30 years. We're plateau-ing out in biomass, and there's no easy answer for it.**
>
> Robert S. Loomis, 1999

and other forces could also reduce yields from present levels, but we will assume in figure 3-4 that yields are maintained or doubled this century.

Assuming current yields are maintained, line A projects the hectares of land required to feed the population at average year-2000 Western European levels. Line B shows land requirements to sustain the present, inadequate diet for the world's population throughout this century. Assuming yields are doubled, line C projects the hectares of land required to feed the global population at average year-2000 Western European levels. Line D shows land requirements to sustain the present, inadequate diet for the world's population throughout this century.

You can see in figure 3-4 how quickly exponential population growth may move the world from land abundance to land scarcity.

But figure 3-4 also shows how many adaptive responses there could be, depending on the resilience of the resource base and the technical and social flexibility of humankind. If no more land were lost, if yields could double worldwide, if degraded land could be restored, then every one of the current six billion people would have enough food, and so would the nearly nine billion projected for the middle of the twenty-first century. But if erosion increases, if irrigation rates cannot be maintained, if developing or restoring land proves too expensive, if another global-average doubling of yield is too

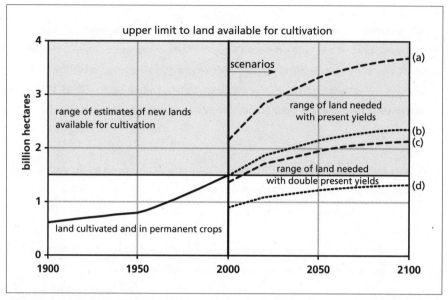

FIGURE 3-4 Possible Agricultural Land Futures

Cultivatable land in the twenty-first century will probably be in the range of 1.5 to 4.0 billion hectares, shown above by the shaded band. Assume here that population growth follows the median UN projection. Scenarios after the year 2000 show the range of land required for food production with current yields per hectare and with double current yields for maintaining present dietary standards and for achieving a global average nutrition equal to that of a typical Western European in the year 2000. (Sources: UN, FRB, FAO, G.M. Higgins et al.)

difficult or environmentally hazardous, if population growth does not level off according to the UN forecast, food could become severely limiting, not only locally, but also globally—and quickly. The scarcity would look sudden, but it would be nothing more than a continuation of exponential trends.

Unsustainable use of the agricultural resource base is a consequence of many factors, including poverty and desperation, expansion of human settlements, overgrazing and overcropping, ignorance, economic rewards for short-term production rather than long-term stewardship, and managers who understand too little about ecology, especially soil ecosystems.

There are other limits to food production besides soils and land, among them water (which we will come to in a minute), energy, and sources and sinks for agricultural chemicals.[20] In some parts of the world, some of these limits have already been surpassed. Soils are eroding, irrigation is drawing down water tables, runoff from agricultural fields is polluting surface and groundwater. For example, the world's large water bodies contain 61 major dead zones—areas where nutritive runoff, mainly from fertilizers and soil erosion, has killed virtually all aquatic life. Some occur year-round, others only in summer, after spring runoff washes down fertilizer residues from upriver farmlands. The Mississippi dead zone covers 21,000 square kilometers (8,000 square miles), the size of the state of Massachusetts.[21] Agricultural practices that produce ecological disturbance at such huge scales are not sustainable. Nor are they necessary.

In many areas soil is not eroding, land is not being abandoned, and agricultural chemicals do not pollute land and water. Farming methods that conserve and enhance soils—such as terracing, contour plowing, composting, cover cropping, polyculture, and crop rotation—have been known and used for centuries. Other methods particularly applicable in the tropics, such as alley cropping and agroforestry, are being demonstrated in experiment stations and on farms.[22] On farms of all types, in both temperate and tropic zones, high yields are obtained sustainably without large applications of synthetic fertilizers and pesticides, often without *any* synthetic fertilizers or pesticides.

Note that the previous sentence said *high yields*. It is a well-documented fact that "organic" farmers need not be primitive or retreat to the agricultural methods and low productivity of 100 years ago. Most of them use high-yielding varieties, labor-saving machines, and sophisticated ecological methods of fertilization and pest control. Their yields tend to be equivalent to those of their chemical-using neighbors; their profits tend to be higher.[23] If even a fraction of the research devoted to chemical inputs and genetic modification were devoted to organic production methods, organic farming would be even more productive.

> In comparison with conventional, high-intensity agricultural methods, "organic" alternatives can improve soil fertility and have fewer detrimental effects on the environment. These alternatives can also produce equivalent crop yields to conventional methods.[24]

Sustainable agriculture not only is possible, but is already practiced in some places. Millions of farmers in all parts of the world employ ecologically sound agricultural techniques, discovering that as soil degradation is reversed, yields continue to rise. Consumers, in the rich world at least, increasingly demand food produced this way, and they are willing to pay a premium price for it. In the United States and Europe the market for organic produce grew through the 1990s at 20 to 30 percent per year. By 1998 organic food and beverage sales in the major world markets totaled $13 billion.[25]

Why have we not mentioned the promise of genetically modified crops? Because the jury is still out on this technology—indeed the jury is in deep controversy. It is not clear either that genetic engineering is needed to feed the world or that it is sustainable. People are not hungry because there is too little food to buy; they are hungry because they cannot afford to buy food. Producing greater amounts of high-cost food will not help them. And while genetic engineering might increase yields, there are plenty of still-unrealized opportunities to raise yields without genomic interventions that are both high-tech (therefore inaccessible to the ordinary farmer) and ecologically risky. The rush to biotech crops is already producing troubling ecological, agricultural, and consumer backlashes.[26]

Everyone could be more than adequately nourished with the amount of food now grown. And more food could be grown. It could be done with much less pollution, on less land, using less fossil energy—allowing millions of hectares to be returned to nature or to fiber, forage, or energy production. It could be done in ways that adequately reward farmers for feeding the world. But so far the political will to accomplish those results has been mainly lacking. The present reality is that in many parts of the world the soil, land, and nutrient sources of food are declining, and so are agricultural economies and communities. In those places, given present practices, agricultural production has overshot many kinds of limits. Unless rapid changes are made—changes that are entirely possible—the growing human population will have to try to feed itself with fewer farmers, working with a declining agricultural resource base.

Water

In many countries, both developing and developed, current pathways for water use are often not sustainable. . . . The world faces a worsening series of local and regional water quantity and quality problems. . . . Water resource constraints and water degradation are weakening one of the resource bases on which human society is built.

—UN COMPREHENSIVE ASSESSMENT OF
THE FRESHWATER RESOURCES, 1997

Fresh water is not a global resource. It is a regional one, available within specific watersheds, so limits take many different forms. In some watersheds the limits are seasonal, dependent on the ability to store water through dry periods. In other places limits are determined by groundwater recharge rates, or snowmelt rates, or the water-storing capacity of forest soils. Since water is not only a source but also a sink, its uses may also be limited by the degree to which it has been polluted upstream or underground.

The inherently regional nature of water does not prevent people from making global statements about it—statements that increasingly reflect deep concern. Water is the least substitutable and most essential resource. Its limits constrain other necessary throughputs—food, energy, fish, and wildlife. The extraction of other throughputs—food, minerals, and forest products—can further limit the quantity or quality of water. In an increasing number of the world's watersheds, limits have already, indisputably, been exceeded. In some of the poorest and richest economies, per capita water withdrawals are going down because of environmental concerns, rising costs, or scarcity.

Figure 3-5 is only illustrative, because it is a global summary of many regional watersheds. We could draw a similar graph for every region, however, with the same general characteristics—a limit, a number of factors that can expand or contract the limit, and growth toward—and in some places beyond—the limit.

At the top of the graph is the upper physical limit to human water use, the total annual flow of the streams and rivers of the world (including the recharge of all groundwater aquifers). This is the renewable resource from which nearly all freshwater inputs to the human economy are taken. It is a huge amount of water: 40,700 cubic kilometers per year, enough to fill the five Great Lakes of North America every four months. It would seem to be a far-off limit indeed, given current human water withdrawals of just over 5 percent of that amount: 2,290 cubic kilometers per year.[27]

In practice, however, all that freshwater runoff cannot be used. Much of it is seasonal. As much as 29,000 cubic kilometers per year flows to the sea in floods. That leaves only 11,000 cubic kilometers that can be counted on as a year-round resource, the sum of base river and groundwater recharge flows.

Figure 3-5 shows that humans are raising that runoff limit by building dams to trap floodwater. By the end of the twentieth century, dams had

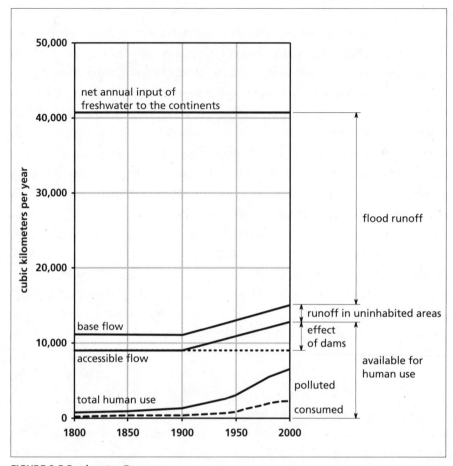

FIGURE 3-5 Freshwater Resources

A graph of global freshwater supply and use shows how quickly growth in consumption and pollution can approach the total amount of water that is accessible—as well as how essential dam construction has been for stability of supply. (Sources: P. Glick; S. L. Postel et al.; D. J. Bogue; UN.)

increased useable runoff by about 3,500 cubic kilometers per year.[28] (Dams flood land, of course, often prime agricultural land. And they generate electricity. They also increase evaporation from the river basin, reducing net runoff and altering both streamside and aquatic ecosystems. Sooner or later they silt up and become ineffective, so they are not a source of sustainable flow; they produce another very long-term delay—with many positive and negative side effects—in feedback from the limits.)

There are other ways besides dams to raise the water limit, such as desalination of seawater or long-distance water transport. These changes can be important locally, but they are energy-intensive and expensive. So far they are too small to show up on a global-scale graph.[29]

Not all the sustainable flow occurs where people live. The Amazon basin accounts for 15 percent of global runoff but is home to only 0.4 percent of the global population. The far northern rivers of North America and Eurasia carry 1,800 cubic kilometers per year in regions where very few people live. Stable runoff not easily accessible to humans amounts to roughly 2,100 cubic kilometers per year.

The 11,000 cubic kilometers in sustainable flow, plus 3,500 added by dams, minus 2,100 that is inaccessible, leaves 12,400 cubic kilometers per year of accessible sustainable flow. That is the foreseeable upper limit to the renewable freshwater supply available for human use.[30]

Human consumptive withdrawal (water removed but not returned to streams or groundwater because it evaporates or is incorporated into crops or products) amounts to 2,290 cubic kilometers per year. Another 4,490 cubic kilometers are used primarily to dilute and carry away pollution. Those two categories of impact add up to 6,780 cubic kilometers per year, just over half the total sustainable freshwater runoff.

Does that mean there is room for another doubling of water use? Is there likely to be another doubling?

If average per capita demand did not change at all and the human population grew to nine billion by the year 2050, as the UN presently projects, humans would withdraw 10,200 cubic kilometers per year, 82 percent of the global sustainable freshwater runoff. If not only population but also per capita demand increased, there would be severe global water limits long before the year 2100. Throughout the 20th century, water withdrawals rose roughly twice as fast as population.[31] But with increasing scarcity, it is likely that per capita consumption will level off and even fall. The withdrawal curve is already beginning to slow markedly and in some places even turn downward. Worldwide water use is only half of what it was predicted to be on the basis of extrapolating exponential curves 30 years ago.[32]

After doubling roughly every 20 years throughout the twentieth century, U.S. water withdrawals peaked around 1980 and have since fallen by about

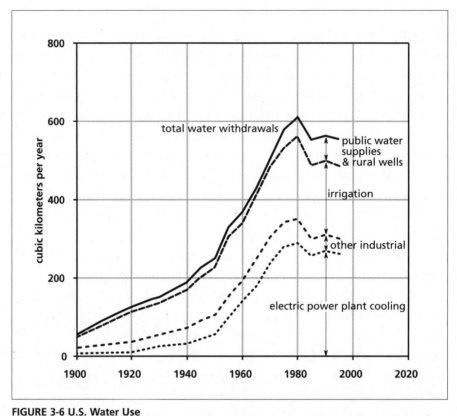

FIGURE 3-6 U.S. Water Use
Water withdrawals in the U.S. grew from the turn of the twentieth century to the 1980s at an average rate of 3 percent per year. Since then withdrawals have dropped slightly and leveled off. (Source: P. Gleick.)

10 percent (figure 3-6). The reasons for this drop are many, all of them relevant to the question of what happens when economies begin to encounter water limits. Industrial use dropped 40 percent, partly because of the export of heavy industry to other parts of the world, but also because of water quality regulations, which made efficient use, recycling, and treatment before release economically attractive or legally mandated (or both). Irrigation use dropped partly because of increased efficiency, and also because expanding municipalities bought water away from farmers (thereby taking land out of food production). Municipal use increased, but only through population growth. Per capita consumption dropped, especially in the arid parts of the nation, where rising water prices encouraged more water-efficient appliances.[33]

U.S. per capita water withdrawal may have declined, but it is still very high at 1,500 cubic meters per person per year. The average citizen of the developing world uses only one-third that much; the average sub-Saharan African, only one-tenth.[34] One billion people still lack access to safe drinking water. Half the human population does not have basic sanitation facilities.[35] Their water demand is certainly going to, and should, go up. Unfortunately, they live in some of the world's most water-deficient places.

> About one-third of the world's population lives in countries that are experiencing moderate to high water stress partly resulting from increasing demands from a growing population and human activities. By 2025 as much as $2/3$ of the world population would be under stress conditions. Water shortages and pollution are causing widespread public health problems, limiting economic and agricultural development, and harming a wide range of ecosystems. They may put global food supplies in jeopardy and lead to economic stagnation in many areas of the world.[36]

The Colorado, Yellow, Nile, Ganges, Indus, Chao Phraya, Syr Darya, and Amu Darya Rivers are so diverted by withdrawals for irrigation and cities that their channels run dry for some or all of the year. In India's agricultural states of Punjab and Haryana, water tables are dropping by half a meter per year. North China overpumps its wells by 30 cubic kilometers a year (which is one reason the Yellow River is running dry). The Ogalalla aquifer that waters one-fifth of the irrigated land in the United States is overdrawn by 12 cubic kilometers per year. Its depletion has so far caused one million hectares of farmland to be taken out of irrigation. California's Central Valley, which grows half the nation's fruits and vegetables, averages a groundwater overdraft of about one cubic kilometer per year. Throughout North Africa and the Middle East, water is being pumped from desert aquifers that have little or no recharge.[37]

Pumping up groundwater faster than it can be recharged is unsustainable. The human activities that depend on it will either have to decline to a level

---

Groundwater overdraft is accelerating. Unsustainable groundwater use occurs on every continent except Antarctica.

Peter Gleick, *The World's Water 1998–99*

that the renewable recharge rate can sustain, or, if the overpumping destroys the aquifer by saltwater infiltration or land subsidence, cease altogether. Initially these responses to water shortages have mainly local effects. But as they are forced on more and more countries, the consequences are felt internationally. The first symptoms of this probably are higher grain prices.

> Water-scarce countries often satisfy the growing needs of cities and industry by diverting water from irrigation and importing grain to offset the resulting loss of production. Since a ton of grain equals 1,000 tons of water, importing grain is the most efficient way to import water. . . . Although military conflicts over water are always a possibility, future competition for water seems more likely to take place in world grain markets. . . . Iran and Egypt . . . now import more wheat than Japan, traditionally the world's leading importer. Imports supply 40 percent or more of the total consumption of grain . . . in both countries. . . . Numerous other water-short countries also import much of their grain. Morocco brings in half its grain. For Algeria and Saudi Arabia the figure is over 70 percent. Yemen imports nearly 80 percent of its grain and Israel, more than 90 percent. . . . China soon will be forced to turn to the world grain market.[38]

The consequences for a society that overshoots its water limit depend on whether the society is rich or poor, whether it has neighbors with water excess, and whether it gets along with those neighbors. Rich societies can import grain. Rich societies with willing neighbors, such as southern California, can construct canals, pipelines, and pumps to import water. (Though in that case some of the neighbors are beginning to ask for their water back.) Rich societies with large oil reserves, such as Saudi Arabia, can use fossil energy to desalinate seawater. (While the fossil fuels last.) Rich societies with neither, such as Israel, can come up with ingenious technologies to use every drop of water with maximum efficiency and shift their economies toward the least water-intensive activities. Some nations can use military force to expropriate or assure access to their neighbors' water resources. Societies with none of those advantages must develop severe rationing and regulation schemes, or experience famine and/or internal conflict over water.[39]

As with food, there are numerous ways to work toward water sustainability, not by trying to produce more, but by making much more effective use of less. Just a short list of possibilities includes:[40]

- Match water quality to use. For example, flush toilets or water lawns with gray water from sink drains, rather than with drinking water.
- Use drip irrigation, which can cut water use by 30 to 70 percent while raising yields by 20 to 90 percent.
- Install low-flow faucets, toilets, and washing machines. U.S. average household use of 0.3 cubic meter per person per day could be cut in half with efficient water-using appliances—which are both available and affordable.
- Fix leaks. It's amazing how many municipal water authorities spend money on increasing water supply when, for a fraction of the cost, they could obtain as much water by fixing leaks. The average U.S. city loses about a quarter of its piped water to leaks.
- Plant appropriately to the climate, such as by not growing water-intensive crops like alfalfa or corn in the desert, and by landscaping with native plants that do not require watering.
- Recycle water. Some industries, particularly in water-scarce California, have pioneered efficient, cost-effective techniques for recapturing, purifying, and reusing water.
- Collect rainwater in urban areas. A cistern or collection system from roofs can store and make use of as much runoff water as a major dam at much lower cost.

One of the best ways to put these good practices into action is to stop subsidizing water. If water price began to incorporate even partially the full financial, social, and environmental cost of delivering that water, wiser use would become automatic. Both Denver and New York discovered that just metering city water, with a charge that rises with rate of use, reduced household use by 30 to 40 percent.

And then there's climate change (about which more below). If humanity lets it progress, it could change the hydrological cycle, the ocean currents, the precipitation and runoff patterns, the efficacy of dams and irrigation systems, and other forms of water storage and delivery capital everywhere on earth. Water sustainability is not possible without climate sustainability, which means energy sustainability. Humanity is dealing with one, large, interlinked system.

Forests

There has been a clear global trend toward a massive loss of forested areas. . . . The current trends are toward an acceleration of the loss of forested area, the loss of residual primary forests, and progressive reduction in the internal quality of residual forest stands. . . . Much of the forest that remains is being progressively impoverished, and all is threatened.

—WORLD COMMISSION ON FORESTS AND
SUSTAINABLE DEVELOPMENT, 1999

A standing forest is a resource in itself, performing vital functions beyond economic measure. Forests moderate climate, control floods, and store water against drought. They cushion the erosive effects of rainfall, build and hold soil on slopes, and keep rivers and seacoasts, irrigation canals and dam reservoirs, free from silt. They harbor and support many species of life. The tropical forests alone, which cover only 7 percent of the earth's surface, are believed to be the home of at least 50 percent of the earth's species. Many of those species, from rattan vines to mushrooms to sources of medicines and dyes and food, have commercial value, and cannot exist without the sheltering trees that form their habitat.

Forests take in and hold a great stock of carbon, which helps balance the stock of carbon dioxide in the atmosphere and thus ameliorates the greenhouse effect and global warming. And last, but far from least, undisturbed forests are beautiful, beloved places for recreation and the restoration of the human soul.

Before the advent of human agriculture, there were six to seven billion hectares of forest on Earth. Now there are only 3.9 billion, if we include some 0.2 billion hectares of forest plantations. More than half the loss of the world's natural forests has occurred since 1950. Between 1990 and 2000 the area of natural forests decreased by 160 million hectares, or about 4 percent.[41] Most of the loss was in the tropics: The destruction of the temperate forests took place long before 1990 during the industrialization of Europe and North America.

Loss of forest is an obvious sign of unsustainability—a renewable resource stock is shrinking. But as is often the case, beneath the clear global trend lie complicated local differences.

We need to distinguish between two measures of the forest resource— area and quality. There is a huge difference between a hectare of undisturbed forest with trees hundreds of years old and a resprouting clear-cut that won't have an economically valuable tree on it for 50 years and may never again have the ecological diversity of a primary forest. Nevertheless, many nations' data on forest area do not differentiate between the two.

Forest quality is much harder to measure than forest area. The least disputed quality data are actually area-related; they are the statistics on the remaining area of forests that have never been cut (called primary, frontier, or old-growth forests). There is no doubt that these valuable forests are being rapidly converted into lower-value ones.

Only one-fifth (1.3 billion hectares) of the Earth's original forest cover remains in large tracts of relatively undisturbed natural forests.[42] Half of this is boreal forest in Russia, Canada, and Alaska; much of the rest is tropical rain forest in the Amazon. Vast areas are threatened by logging claims, mines, agricultural clearing, and other human activity. Only about 0.3 billion hectares are formally protected (and some of this protection is only on paper; in many of these forests, the wood and/or the wildlife are systematically poached).

The United States (exclusive of Alaska) has lost 95 percent of its original forest cover. Europe has essentially no primary forest left. China has lost three-fourths of its forest and nearly all its frontier forest. (See figure 3-7.) Logged but regrown (secondary) temperate-zone forests are increasing slightly in area, but many are declining in soil nutrients, species composition, tree size, wood quality, and growth rate; they are not being managed sustainably.

Less than half of the remaining natural forest is found in temperate

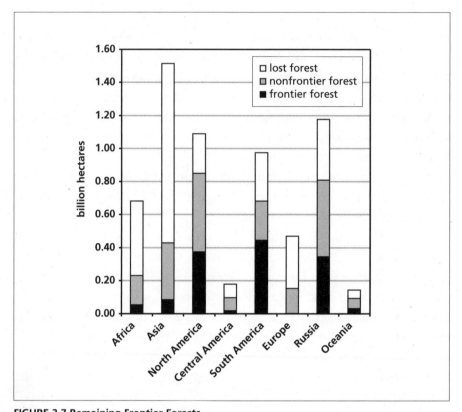

FIGURE 3-7 Remaining Frontier Forests
Only a small fraction of the original global forest cover remained in 1997 as untouched "frontier" forests. (Source: WRI.)

zones (1.6 billion hectares); the rest is in the tropics (2.1 billion hectares). Between 1990 and 2000 the area of natural forest in the temperate zone declined only slightly, by some nine million hectares, which equals around 0.6 percent lost during the decade. Half of this natural forest was converted to intensively managed tree plantations to supply paper or lumber. In addition, about the same acreage was reforested.

While temperate forest area is essentially stable, the tropical forest area is plummeting. From 1990 to 2000, the FAO reports that more than 150 million hectares of the world's remaining natural tropical forests—an area similar to that of Mexico—were converted to other uses. Thus, in the 1990s the loss seems to have been around 15 million hectares per year, or 7 percent during the decade.

That is the official number, but no one is sure exactly how fast the tropical forest is being cleared. The numbers change from year to year and are in dispute. That in itself—the fact that the loss rate of the resource is unclear—is one of the structural causes of overshoot.

The first authoritative attempt to assess tropical deforestation rates, conducted by the FAO in 1980, came up with a figure of 11.4 million hectares lost per year. By the mid-1980s the estimated rate had climbed to more than 20 million hectares per year. After some policy changes, particularly in Brazil, the rate of loss by 1990 had apparently come down to around 14 million hectares per year. A new FAO assessment in 1999 set the annual rate of forest loss, nearly all of it in the tropics, at 11.3 million hectares per year. And, as noted above, at the end of the decade the final estimate came out at 15.2 million hectares per year.

That number counts only permanent conversion to other forms of land use (mainly agriculture and grazing; secondarily roads and settlements). *It does not count logging* (since a logged forest is still counted as a forest). And it does not count fires, which burned 2 million hectares in Brazil, 2 million hectares in Indonesia, and 1.5 million hectares in Mexico and Central America in 1997–98. (Burned-over land is still classified as forest.) If we add in the net rate at which the area labeled tropical forest is becoming treeless, the total almost certainly exceeds 15 million hectares per year, and might approach 1 percent of the forested area per year.

Despite the uncertainty in the data, we can use rough numbers to get an idea of the likely fate of the natural tropical forest if there are no changes to the present system. Figure 3-8 starts with the estimated year-2000 total tropical forest area of 2.1 billion hectares. We assume that the current loss rate is 20 million hectares per year, higher than the official FAO estimate to correct for fires, unsustainable logging, and underreporting. The horizontal line in the graph represents the limit to forest removal if 10 percent of the present-day tropical forests remain protected. (This is roughly the percentage of the tropical forest estate that is currently under some form of protection.[43])

If the clearing rate stays constant at 20 million hectares per year, the unprotected primary forest will be gone in 95 years. This possibility is shown by the straight line in figure 3-8. It reflects the situation where the forces that cause forest destruction will neither strengthen nor weaken over the coming century.

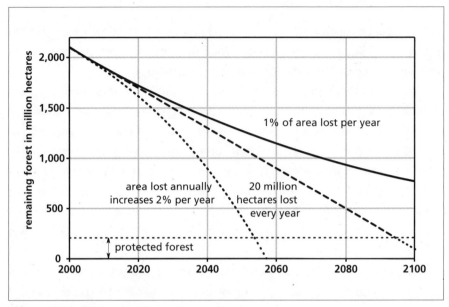

FIGURE 3-8 Some Possible Paths of Tropical Deforestation
Estimates of the future loss of tropical forests depend upon assumptions about demographic, legal, and economic trends. Three scenarios are shown in this graph. If the loss of 20 million hectares per year, which was typical in the 1990s, continues to increase at 2 percent per year, the unprotected forest will be gone by 2054. If the rate of loss remains constant at 20 million hectares per year, the unprotected forest will be gone by about 2094. If the rate of loss is 1 percent of the remaining unprotected forest each year, the forest will shrink to half its size every 72 years.

If the clearing rate increases exponentially, say at the rate the population of the tropical countries is growing (about 2 percent per year), the unprotected forest will be completely gone in about 50 years. This curve reflects the situation where some combination of population growth and growth in the forest products industry will force the rate of forest loss to grow exponentially.

If the clearing rate remains a constant percentage of the remaining forest (say, that 1 percent is cleared every year), the cut will be slightly less each year compared with the year before, because there is less forest each year. If this continues, half the tropical forest area will be gone after 72 years. This curve reflects the situation where each cut makes the next one less likely, perhaps because the nearest, most valuable forests are cut first.

The actual future will probably combine all these possibilities. As population and economic growth drive demand for forest products and cleared land upward, increasing remoteness and declining quality will make logging more costly. At the same time, environmental and political pressures will likely increase to protect the remaining forest and shift wood production to high-yield plantations. However these conflicting trends resolve themselves, one outcome seems unavoidable: The present stream of products from primary tropical forests—forests that were planted and tended by nature at no cost to the human economy and that had time to grow trees of great size and value—is unsustainable.

Tropical soils, climates, and ecosystems are very different from temperate ones. Richer in species, tropical forests are faster growing, but also more vulnerable. It is not certain that they can survive even one clear-cut or fire without serious degradation of soil and ecosystem. While experiments are under way to find a method of logging tropical forests selectively or in strips to allow regeneration, most current logging practices treat the tropical forest, especially its most valuable tree species, as a nonrenewable resource.[44]

The reasons for tropical forest loss vary from one country to another. Drivers include multinational timber and paper companies seeking higher sales; governments increasing exports to pay external debts; ranchers and farmers converting forest to agricultural or grazing land; and landless people scrambling for firewood or a patch of land on which to grow food. These actors often work in concert, the government inviting in the companies, the companies harvesting the wood, the poor moving in along logging roads to find a place to settle.

There is another driver of unsustainable forest use in both temperate and tropic zones. In a world of disappearing high-quality lumber, a single old-growth tree can be worth $10,000 or more. That value sets up enormous temptations. Giveaways of publicly owned forest resources for private gain, secret sales of harvesting permits, misleading accounting, false certification of species or volumes or areas cut, halfhearted enforcement of regulations, sweetheart deals, kickbacks—these practices do not take place only in tropical areas.

> The Commission found that the most readily perceived problem in the forest sector—most prevalent and most blatant—but the one least discussed . . . is the existence of corrupt practices.[45]

Even in the least corrupt, most concerned tropical countries, the forest is shrinking, but it's not easy to know at what rate. In the 1992 edition of this book, we showed maps of forest loss in one small country, Costa Rica. Seeking to update this figure, we contacted the Research Center for Sustainable Development at the University of Costa Rica, only to learn that data on earlier years had had to be revised as better measurement techniques became available.

Compounding the problem of forest decline, demand for forest products is growing. Between 1950 and 1996 world paper consumption grew by a factor of six. FAO expects it to rise from 280 to 400 million tons by 2010.[46] In the United States the average person uses 330 kilograms of paper per year. In the other industrial nations the average person uses 160 kilograms; in the developing world, just 17. Though paper recycling is increasing, the consumption of virgin wood for pulp continues to go up by 1 to 2 percent per year.

Total consumption of wood for all purposes—construction timber, paper products, and fuel wood—is growing, though the rate of growth is slowing (figure 3-9). One reason for the slower growth rate in the 1990s appears to be the decline in the Asian and Russian economies. The leveling off of roundwood consumption may thus be a temporary phenomenon. If everyone in the world consumed as much wood for all purposes as the average person in industrial countries does today, total wood consumption would more than double.[47]

There are also trends toward reducing wood demand, however, such as recycling and using forest products more efficiently. If these trends increase, the world could easily meet its need for wood products with a much smaller throughput stream from the forests. For example:

- *Paper recycling.* About half the paper made in the United States comes from recycled paper; in Japan the figure is more than 50 percent, and in Holland, 96 percent. Worldwide, 41 percent of paper and paperboard is recycled.[48] If the world could emulate Holland, the paper recycling rate could more than double.
- *Mill efficiency.* Modern sawmills turn 40 to 50 percent of their entering logs into salable lumber (and the residue into fuel or paper or composite lumber made from glued-together chips). Less efficient mills, especially in the developing world, make use

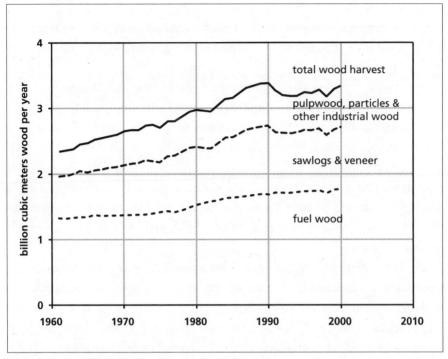

FIGURE 3-9 World Wood Use
The use of wood is still growing, albeit at a slower rate. Roughly half of the wood removed from the world's forests is used for fuel. (Source: FAO.)

of only 25 to 30 percent of each log. If these inefficient mills could be updated, twice as much lumber could be produced per tree cut down.[49]

- *Fuel efficiency.* More than half the wood cut from forests is used for cooking and heating and small industries (brick making, brewing, tobacco curing) by the poor, often in extremely inefficient wood-burning stoves or open fires. Greater stove efficiency or alternative fuels could supply human need with much less forest consumption, less air pollution, and less labor to gather the fuel.

- *Paper use efficiency.* Half the world's paper and paperboard production is used for packaging and advertising. The average U.S. household receives 550 pieces of unsolicited "junk mail" per year, most of which is discarded without being read. Despite the electronic age, or maybe because of it, per capita paper consumption in the United States doubled from 1965 to 1995. Junk mail and overpackaging could be eliminated; single-side-only laser printers and fax machines and many other wasteful technologies could be improved.

- *Full-cost pricing.* Direct and indirect government subsidies to logging could be removed, and logging taxes reflecting the lost value of standing timber could be imposed, so the prices of wood products would signal more realistically their actual cost.

Advances like these in the industrial countries could probably cut their throughput stream of wood from the forests—and the waste stream at the discard end—by a factor of at least two, with little if any sacrifice in the quality of life.

At the same time, the valuable fibers from the forest could be produced with far less damage. Clear-cutting, especially on steep slopes, could be replaced by selective or strip cutting. Uncut buffer strips along streams would reduce erosion and protect aquatic ecosystems from damaging sunlight. Some dead trees, standing and fallen, could be left as habitat.

There is a growing movement toward "green certification," which allows consumers to identify forest products produced with careful harvesting and forest management practices. By the end of 2002 the Forest Stewardship Council had certified a total of 30 million hectares of forest as "sustainably managed." Although small, this number is growing quickly, demonstrating the power of the market—in this case the power of consumer demand for certified wood.

High-yield forest plantations could be expanded on already cleared or marginal lands. Tree plantations can produce astonishing amounts of wood per hectare, releasing natural forests from logging pressure.

To take an extreme example, the highest-yielding tropical forest plantations can yield (for a while, anyway) as much as 100 cubic meters of wood per hectare per year. This exceeds by a factor of 40 the average growth rate of natural temperate-zone forests, which yield around 2.5 cubic meters per hectare per year. At the high plantation rate, it would take only 34 million hectares (an area roughly the size of Malaysia) to supply the entire present world demand for virgin pulp, construction wood, and fuel wood. If productivity were only half that high, 50 cubic meters per hectare per year, it would take 68 million hectares (the size of Somalia) to supply present world

demand. To maintain the tremendous productivity of tropical forest plantations sustainably would certainly require a more "organic" evolution of
plantation forestry: mixing or rotating species and using more natural, less
environmentally destructive methods of fertilization and pest control than
are currently in use.

There are many ways to bring forest harvest rates back down below sustainable limits. None of the necessary measures is impossible. Every one of
them is being practiced somewhere in the world, but not in the world as a
whole. And so the forests continue to shrink.

> Although public awareness of the impact of global deforestation has
> increased in recent years, it has not slowed the rate of deforestation
> appreciably.[50]

Species and Ecosystem Services

> The Living Planet Index is an indicator of the state of the
> world's natural ecosystem. It . . . relates to the abundance of
> forest, freshwater, and marine species. The index shows an
> overall decline of about 37 per cent between 1970 and 2000.
> —WORLD WIDE FUND FOR NATURE, 2002

Soils, waters, and forests are obvious sources upon which humans depend
for throughputs that sustain life and economy. There is another set of
sources, at least as important but far less obvious, because the human
economy has never put a monetary value on them. They are the noncommercial, unmarketed natural species, the ecosystems they form, and the
support functions they provide, as they capture, mobilize, and recycle the
energy and materials needed for all life.

The emerging term for the daily, invaluable contributions of these biotic
sources is *ecosystem services*. The services include:

- Purification of air and water.
- Water absorption and storage; mitigation of drought and floods.
- Decomposition, detoxification, and sequestering of wastes.

- Regeneration of soil nutrients; buildup of soil structure.
- Pollination.
- Pest control.
- Seed and nutrient dispersal.
- Moderation of wind and temperature extremes; partial stabilization of climate.
- Provision of a wide variety of agricultural, medicinal, and industrial products.
- Evolution and maintenance of the biotic gene pool and the biodiversity that performs all of the above tasks.
- Lessons in survival, resilience, evolution, and diversification strategies that have proved themselves over three billion years.
- Unparalleled aesthetic, spiritual, and intellectual uplift.[51]

Although the value of these services is beyond measure, people do attempt to measure it. All attempts to value natural services in money terms come up with estimates in the trillions of dollars per year, far exceeding the monetary value of the annual output of the human economy.[52]

The WWF measurement quoted above implies that the world has lost a significant fraction of its ecosystem services over the past 30 years. Still, this is very difficult to document in quantitative terms. The most common approach, though not very meaningful, is to try to count the number of species and the rate of their extinction.

Surprisingly, this is impossible to do. Scientists do not know within a factor of 10 how many species there are: The number is estimated to be anywhere from 3 million to 30 million.[53] Only 1.5 million or so have been named and classified. They tend to be the large, noticeable species: the green plants, the mammals and birds and fish and reptiles. Science knows far less about the myriad species of insects and even less about the microbes.

Since no one knows how many species there are, no one can know precisely how many are being lost. But there is no doubt that the number of species is decreasing rapidly. Most biologists do not hesitate to say that a "mass extinction" is under way.[54] Ecologists claim there has not been such an extinction wave since the events that eliminated the dinosaurs at the end of the Cretaceous age 65 million years ago.

They come to such conclusions primarily from the rate at which habitat is disappearing. For example:

- Madagascar is a biotic treasure house; its eastern forest houses 12,000 known plant species and 190,000 known animal species, at least 60 percent of which are found nowhere else on Earth. More than 90 percent of that forest has been cleared, primarily for agriculture.
- Western Ecuador once contained 8,000 to 10,000 plant species, about half of them endemic. Each species of plant supports between 10 and 30 animal species. Since 1960 nearly all the western Ecuadorian forests have been turned into banana plantations, oil wells, and human settlements.

Most extinctions happen, as you might expect, where the most species are located. These are principally in tropical forests, coral reefs, and wetlands. At least 30 percent of coral reefs worldwide are in critical condition, and 95 percent of those checked around the world in 1997 showed degradation and species loss.[55] Wetlands are even more endangered. They are places of intense biological activity, including the breeding of many species of fish. Only 6 percent of the earth's surface is wetland—or was. About half the wetlands have been lost to dredging, filling, draining, and ditching. That doesn't count what has been degraded by pollution.

Estimates of global extinction rates start with measures of habitat loss, which are fairly accurate. They go on to assume how many species might be in the habitat that is lost; those assumptions are necessarily uncertain. Then they assume a relationship between habitat loss and species loss. The rule of thumb is that 50 percent of the species will remain even if 90 percent of the habitat is gone.

These calculations are subject to considerable argument.[56] But as with other numbers we are trying to grapple with in this chapter, their general direction is clear. Of the large animals that are relatively well studied, scientists now estimate that 24 percent of the world's 4,700 mammal species, an estimated 30 percent of the 25,000 fish species, and 12 percent of the world's nearly 10,000 bird species are in danger of extinction.[57] The same is true for

34,000 of the 270,000 known species of plants.[58] The estimated rates of extinction are now 1,000 times what they would be without human impact.[59]

Species loss is not a satisfying way of measuring the sustainability of the biosphere, because no one knows where the limits are. How many species, and which ones, can be removed from an ecosystem before the whole system collapses? The comparison has been made to riding in an airplane and removing the rivets that hold it together—one by one—wondering how many rivets can go before the airplane stops flying. At least in an airplane the rivets are not connected to each other. In ecosystems, the species are. If one goes, it may take others with it in a long chain of reactions.

Given the difficulty of measuring the rate at which the number of species on the planet is dwindling, the WWF in its *Living Planet Index* has chosen a different method for quantifying the decline in the biological wealth. Instead of tracking the decline in the number of species, the WWF tracks the population sizes of a large number of different species. These trends are then averaged to obtain a quantitative estimate of the change over time in the population of a "typical" species. By this method the WWF has concluded that the "average" species population has declined by more than one-third since 1970.[60] In other words, the number of animals, plants, and fish is in steep decline. Clearly, the source of ecosystem services is being used unsustainably. This point was made strongly by a 1992 appeal, "World Scientists' Warning to Humanity." It was issued by about 1,700 of the world's leading scientists, including the majority of Nobel laureates in the sciences.

> Our massive tampering with the world's interdependent web of life—coupled with the environmental damage inflicted by deforestation, species loss, and climate change—could trigger widespread adverse effects, including unpredictable collapses of critical biological systems whose interactions and dynamics we only imperfectly understand. Uncertainty over the extent of these effects cannot excuse complacency or delay in facing the threats.[61]

Nonrenewable Sources

Fossil Fuels

Our analysis of the discovery and production of oil fields around the world suggests that within the next decade, the supply of conventional oil will be unable to keep up with demand. . . . Global [oil] discovery peaked in the early 1960s and has been falling steadily ever since. . . . There is only so much crude oil in the world, and the industry has found about 90 percent of it.

—Colin J. Campbell and Jean H. Laherrère, 1998

At present, there is little near-term concern over petroleum supplies . . . The world's petroleum resources are finite, however, and global production will eventually peak and then start to decline. . . . More conventional estimates suggest that global production will not peak for another decade or two, somewhere between 2010 and 2025.

—World Resources, 1997

Optimists and pessimists differ by a few decades in the timing of its production peak. But there is substantial consensus that petroleum is the most limited of the important fossil fuels, and it's global production will reach a maximum sometime during the first half of this century. The human economy's annual energy use grew by an average of 3.5 percent per year from 1950 to 2000. World energy consumption has climbed unevenly but inexorably through wars, recessions, price instabilities, and technical changes (figure 3-10). Most of that energy is used in the industrialized world. The average Western European uses 5.5 times as much commercial energy[62] as the average African. The average North American uses nine times as much as the average Indian.[63] But that is commercial energy, which many must do without:

More than a quarter of the world's population has no access to electricity, and two-fifths still rely mainly on traditional biomass for their

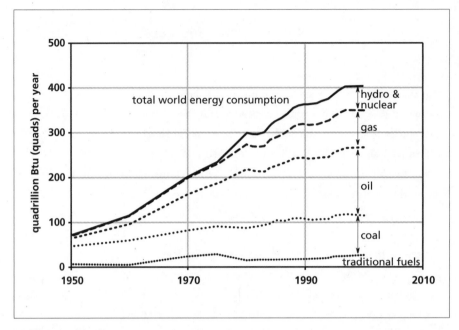

FIGURE 3-10 World Energy Use
World energy use doubled three times between 1950 and 2000. Fossil fuels still dominate the primary energy supply: coal's share peaked around 1920, when it provided more than 70 percent of all fuel consumed; oil's share peaked in the early 1980s at slightly more than 40 percent. Natural gas, which is less polluting than either coal or oil, is expected to contribute more in the future to global energy use. (Sources: UN; U.S. DoE.)

basic energy needs. Although the number of people without power supplies will fall in the coming decades, a projected 1.4 billion people will still be without electricity in 2030. And the number of people using wood, crop residues and animal waste as their main cooking and heating fuels will actually grow.[64]

Most energy analysts expect world energy use to continue rising. The "reference" scenario presented by the International Energy Agency in its *World Energy Outlook 2002*, quoted above, describes an increase in the global primary energy consumption by two-thirds from 2000 to 2030. And even the "alternative" (more ecological) scenario implies more than a 50 percent increase in world energy consumption over this 30-year period. A detailed analysis for the Danish Energy Agency calculates that to supply 9.3 billion people—which may well be the global population in 2050—with all their basic energy needs would require six times as much delivered (end-use) energy as the world supplied in 2000.[65]

More than 80 percent of year-2000 commercial energy use comes from the nonrenewable fossil fuels: oil, natural gas, and coal. The underground stocks of these fossil fuels are going continuously and inexorably down. To determine whether that is a sustainability problem on the source side of the flow (we'll come to the sink side later), we need to ask how rapidly these sources are being depleted and whether renewable substitutes are being developed fast enough to compensate for the decline.

There is tremendous confusion about this matter, even about whether these inherently nonrenewable fuels are in fact being depleted at all. The confusion comes from paying attention to the wrong signal. *Resource* is a concept related to the total quantity of a material in the crust of the earth; *reserve* is a concept related to the amount of the material that has been discovered or inferred to exist and that can be used, given reasonable assumptions about technology and price. Resources go inexhorably down with use, but reserve figures may go up, as discovery proceeds, prices rise, and technology improves. There has been a tendency to make statements about resources based on observations about reserves.

Between 1970 and 2000 the world economy burned 700 billion barrels of oil, 87 billion tons of coal, and 1,800 trillion cubic feet of natural gas. Over that same 30-year period, however, new deposits of oil, coal, and gas were discovered (and old ones were reappraised upward). As a consequence, the ratio of known reserves to production[66]—the number of years known and exploitable resources will last if production continues at its current rate—actually went up, as shown in table 3-1.

This increase in reserve–production ratios occurred despite the significant increase in the consumption of gas (up about 130 percent from 1970 to 2000), oil (up about 60 percent), and coal (up about 145 percent). But does this increase mean there were more fossil fuels left in the ground to power the human economy in 2000 than there were in 1970?

No, of course not. After three decades of exploitation, there were 700 *fewer* billion barrels of oil, 87 billion *fewer* tons of coal, and 1,800 trillion *fewer* cubic feet of natural gas. Fossil fuels are nonrenewable resources. When they are burned, they turn into carbon dioxide, water vapor, sulfur dioxide, and a number of other substances, which do not, on any time scale meaningful to humanity, come back together to form fossil fuels again. Rather, they are wastes and pollutants that enter planetary sinks.

TABLE 3-1 Annual Production, Reserve/Production (R/P) Ratios, and Resource Life Expectancy for Oil, Gas, and Coal

	1970 Production (per year)	1970 R/P (years)	2000 Production (per year)	2000 R/P (years)	Resource Life Expectancy (years)
OIL	17 billion barrels	32	28 billion barrels	37	50–80
GAS	38 trillion cu. ft.	39	88 trillion cu. ft.	65	160–310
COAL	2.2 billion tons	2300	5.0 billion tons	217	very large

The estimates for resources are defined as the sum of "identified reserves" and "conventional resources remaining to be discovered." A resource divided by 2000 production yields 2000 life expectancy for that resource. The reserve figure for coal for 1970 is not comparable to the 2000 figure because of different definitions of reserves. Coal was and is still the most abundant fossil fuel. (Sources: U.S. Bureau of Mines; U.S. DoE)

Those who see the discoveries of the past 30 years as an indication that there are no imminent limits to fossil fuels are looking at only part of the energy system:

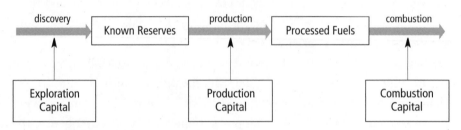

Known Reserves to Processed Fuels

The process of *discovery* uses *exploration capital* (drilling rigs, airplanes, satellites, a sophisticated array of sounders and probes) to find fossil fuel deposits in the earth and thereby to increase the *known reserves* that have been identified but not yet extracted. The process of *production* pulls that stock out of the ground, using *production capital* (mining, pumping, refining and transport equipment), and delivers it to the places where processed fuels are stored. There *combustion capital* (furnaces, automobiles, electricity generators) burns the *processed fuels*, creating useful heat.[67]

As long as the rate of discovery exceeds the rate of production, the stock of known reserves goes up. But the diagram above shows only one part of the system. A more complete diagram would include the ultimate sources and sinks for fossil fuels:

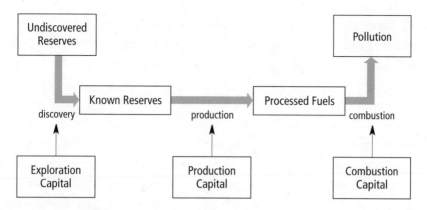

Undiscovered Reserves to Pollution

As *production* reduces the stock of *known reserves*, humankind invests in discovery to replenish it. But every discovery comes from the ultimate stock of fossil fuels in the earth, which is not replenished. The stock of *undiscovered reserves* may be very large, but it is finite and nonrenewable.

At the other end of the flow, combustion produces pollutants, that enter the ultimate sink—the biogeochemical processes of the planet, which recycle pollutants, or render them harmless, or are poisoned or degraded by them. Pollutants of various types are also emitted at every other stage of the fossil fuel flow, from discovery through production, refinement, transportation, and storage. While impressive improvements in eco-efficiency over the past decade have reduced pollution emissions from well-run operations, energy production is still one major source of groundwater pollution in the United States.

No one really knows which end of the flow of fossil fuels will be more limiting, the source or the sink. Thirty years ago, on the eve of the OPEC oil price hikes, the source end seemed the obvious bottleneck. Today the focus is much more on climate change, and therefore the sink end appears more constraining. There is such an enormous amount of coal that we believe its use will be limited by the atmospheric sink for carbon dioxide. Oil may be limited at both ends. Its combustion produces greenhouse gases and other pollutants, and it will certainly be the first fossil fuel to be depleted at the source. Gas is currently viewed by many as the resource that can sustain energy production until there is widespread implementation of sustainable energy sources. But it has traditionally taken society 50 years to make the transition from one dominant energy source to another. Meanwhile the world may suffer welfare loss: either from climatic change, or from limitations on the use of fossil fuels.

Estimates of undiscovered oil and gas reserves vary greatly and can never be certain, but we have included in table 3-1 one set of estimates. These estimates are given as wide ranges because of the inherent uncertainty. They show that remaining oil resources (defined as the sum of current and undiscovered reserves) could last 50 to 80 years at the year 2000 usage rates, while natural gas could last 160 to 310 years. Coal is even more abundant. The cost of accessing the resource will of course increase as the resource depletes. And political costs may add to production costs: in 2000, 30 percent of world oil production came from the Middle East and 11 percent from the former USSR; together these two regions have between them two-thirds of all known oil reserves.

Oil depletion will not appear as a complete stop, a sudden drying-up of the spigot. Rather, it will show up as lower and lower returns on investments in exploration, increasing concentration of the remaining reserves in a few nations, and finally a peak and gradual decline in total world production. The United States provides a case study. Its enormous original oil endowment is more than half gone. Its discovery of new oil peaked in the 1940s and 1950s; its domestic oil production peaked around 1970; its oil consumption is increasingly met by imports (see figure 3-11).

The same is about to happen at the global level. Figure 3-12 shows two scenarios for global oil production, based on resource assumptions similar to those shown in table 3-1. The expectation is that oil consumption will never increase much from present day levels, and then, after some decades, gradually decline during the rest of the twenty-first century. Such scenarios are supported by the fact that global discovery rates did peak already in the 1960s, and that increasingly inaccessible—and thus more expensive—resources are now being exploited, not only in Alaska, but also in the deep waters of the Arctic Ocean and in remote Siberia.

Natural gas is an obvious substitute for oil in many applications. Of all the fossil fuels, natural gas emits the least pollution—including the greenhouse gas CO_2—per energy unit, and therefore there is significant interest in having it rapidly replace oil and coal. This will speed up the depletion of

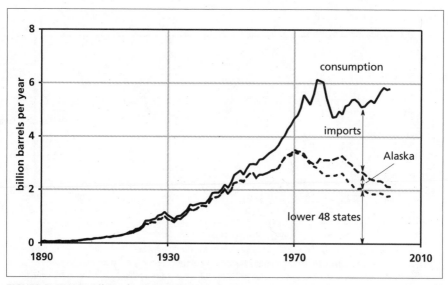

FIGURE 3-11 U.S. Oil Production and Consumption
U.S. domestic oil production peaked in 1970, and production in the lower 48 states has since dropped by 40 percent. Even new discoveries in Alaska have not compensated for the decline. (Sources: API; EIA/DoE.)

gas resources to an extent that will surprise those who do not fully appreciate the dynamics of exponential growth. Figures 3-13 and 3-14 show why.

In 2000 the world reserve–production ratio for natural gas was 65 years, which means that if current known reserves continued to be used at 2000 consumption rates, they would last until the year 2065. Two things will happen to make that simple extrapolation wrong. One is that more reserves will be discovered. The other is that gas use will grow above the 2000 rate.

So it is better to start from the estimates of remaining gas resources (that is, the sum of current and undiscovered reserves). Suppose, for purposes of illustration, that the gas resources in the end will prove sufficient to supply the world at the 2000 usage rate for 260 years. This falls somewhere in the middle of the estimate of 160 to 310 years given in table 3-1. If the 2000 usage rate remained constant, gas resources would go down linearly, as illustrated by the diagonal line in figure 3-13, and they would last 260 years. But if gas consumption continues to grow as it has since 1970, at about 2.8 percent per year, the 260-year resource endowment would plummet exponentially, as shown by the thickest curved line in figure 3-13. It would be exhausted not in 2260, but in 2075; it would last not 260 but only 75 years.

If, to reduce climate change and escape oil depletion, the world calls upon natural gas to carry the energy load now handled by coal and oil, the

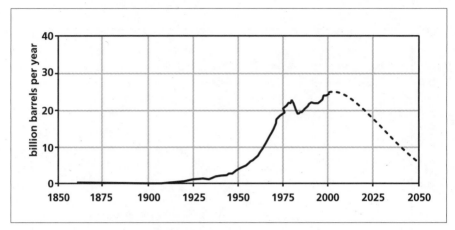

FIGURE 3-12 Scenarios for Global Oil Production
World oil production through the year 2000 is shown by the solid line. Geologist M. King Hubbert's methods were used to estimate the most likely future production. The dashed lines on the right show the probable production rates if the ultimate discovereable oil is 1.8 trillion barrels (the area under the lower curve). (Source: K. S. Deffeyes.)

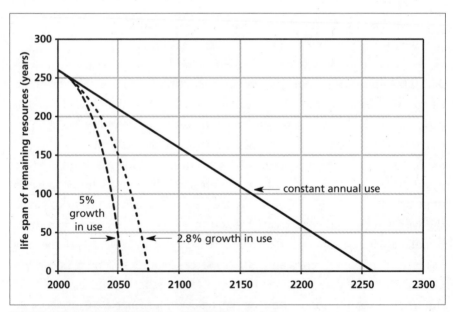

FIGURE 3-13 Some Possible Paths of World Gas Depletion
If the remaining "ultimately recoverable resources" of natural gas can supply 260 years at 2000 usage rates, then this consumption rate can be sustained until 2260. But scarcity of oil combined with coal's environmental problems could accelerate gas use over the decades to come. If gas consumption were to continue to grow at its present rate of 2.8 percent per year, the assumed resource base would be depleted by 2075. At 5 percent per year growth, the world gas resource would be gone by 2054.

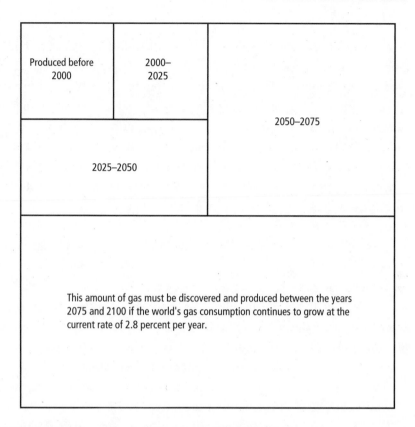

FIGURE 3-14 Gas Discoveries Necessary to Maintain Growth
If the rate of growth of natural gas consumption continues at 2.8 percent per year, every 25 years an amount of new gas must be discovered that is equal to all the previous discoveries.

growth rate could well be faster than 2.8 percent per year. If it were 5 percent per year, the "260-year supply" would be exhausted in 54 years.

Figure 3-14 shows how discoveries would have to increase in order to permit a steady growth of natural gas consumption at 2.8 percent per year. By the mathematics of exponential growth, the amount of gas discovered and extracted would have to double every 25 years.

The point is not that the world is about to run out of natural gas. The considerable resources that remain will be essential as a transition fuel on the way to more sustainable energy sources. The point is that fossil fuels are surprisingly limited, especially when used exponentially, and they should not be wasted. On the time line of human history, the era of fossil fuels will be a short blip.

Because there are renewable substitutes for fossil fuels, there need never be global energy scarcity. Two available energy options are sustainable from the source, environmentally supportable, technically feasible, and increasingly economical. One of them, greater *efficiency*, can be implemented quickly. The other, *solar-based renewables*, will take only a little longer. Some would argue that nuclear energy belongs to the small group of potential solutions to the world's energy problem. We do not think so, because of its unsolved waste processing problems and because the two other solutions are much more feasible. They are quicker, cheaper, safer, and much easier to develop in the poorer nations.

Energy efficiency means producing the same final energy services— light, heat, and cooling rooms, transport for people and freight, pumped water, turning motors—but using less energy to do so. It means the same or better material quality of life, usually at less cost—not only less direct energy cost, but also less pollution, less drawdown of domestic energy sources, less conflict over siting facilities, and, for many countries, less foreign debt and less military cost to maintain access to or control foreign resources.

Efficiency technologies, from better insulation to smarter motors, are improving so quickly that estimates of the energy needed to accomplish any given task have to be revised downward every year. A compact fluorescent lightbulb will give the same amount of light as an incandescent one but use only one-fourth as much electricity. Insulating superwindows in all U.S. buildings could save twice as much energy as the nation now gets from Alaskan oil. At least 10 automobile companies have built prototype cars that drive 30 to 60 kilometers on a liter of gas (65 to 130 miles per gallon), and leading-edge technical discussions are now beginning to speak of 70 km per liter (160 mpg) vehicles. Contrary to popular belief, these efficient cars pass all tests for safety, and some will cost no more to build than current models.[68]

Calculations of how much energy could be saved through efficiency depend on the technical and political biases of the people who do the calculating. On the conservative end of the range, it seems certain that the U.S. economy could do everything it now does, with currently available technologies and at current or lower costs, using half as much energy. That would bring the United States up to the present efficiency levels of Western Europe[69]—and it would reduce the worldwide drain on oil by 14 percent, coal by 14 percent, and gas by 15 percent. Similar or greater efficiency improvements are possible in Eastern Europe and the less industrialized world.

The optimists say that's only the beginning. They believe that Western Europe and Japan, already the most energy-efficient parts of the world, could increase their efficiencies by factors of two to four with technologies already available or easily foreseeable within 20 years. Efficiency of that magnitude would make it possible to supply most or all of the world's energy from solar-based renewable sources—sun, wind, hydropower, and biomass. The sun pours 10,000 times more energy upon the earth every day than human beings currently use.[70]

Technical advances in capturing the sun's energy have been slower than those in raising efficiency, but they have been steady nonetheless. The costs of solar photovoltaic and wind-powered electricity have dropped substantially over the past 20 years (figure 3-15). In 1970 photovoltaic (PV) electricity was generated at a capital cost of $120 per watt. By 2000 the cost had dropped to $3.50 per watt.[71] In less industrialized countries PV is already the most cost-effective choice for villages and irrigation projects that cannot afford the capital cost of connecting to a distant electric grid.

At the costs now attained wind energy has the potential for very rapid growth. At the end of 2002 global installed wind energy capacity exceeded 31,000 MW—the equivalent of more than 30 nuclear power reactors. That represented a 28 percent growth in capacity since the end of 2001 and a fourfold increase in the five years since the end of 1997.[72] Change of this magnitude can encourage all sorts of speculation about the future of energy.

> I believe that we are living through the last days of the traditional oil company. . . . The economics of the world itself change when you park a car and then use its fuel cell to generate electricity for your home. The power grid of an entire country begins to look like the Internet rather than a mainframe. In fact, if all the cars on the road in the U.S. had fuel cells, you would have five times the electrical capacity of today's installed base.[73]

Renewable energy sources are not environmentally harmless, and they are not unlimited. Windmills require land and access roads. Some kinds of

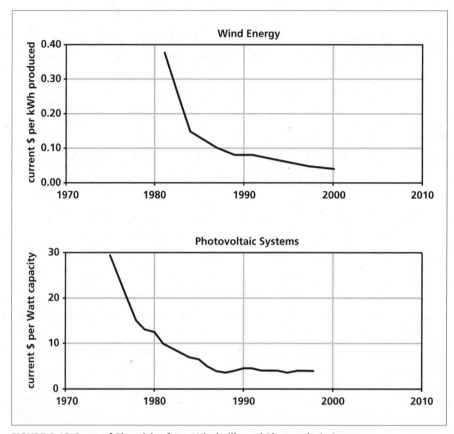

FIGURE 3-15 Costs of Electricity from Windmills and Photovoltaic Systems
Between 1980 and 2000 the cost of electricity generated by windmills and photovoltaic systems fell dramatically. Wind power is now becoming competitive with new fossil-fuel-fired power plants. (Sources: AWEA; EIA/DoE.)

solar cells contain toxic materials. Hydroelectric dams flood land and ruin free-flowing streams. Biomass energy is only as sustainable as the agriculture or forestry practices that produce the biomass. Some solar sources are dilute and intermittent and require large collection areas and complex storage mechanisms,[74] and all require physical capital and careful management. Renewable energy sources are also rate-limited; they can flow forever, but only at a fixed rate. They cannot support an indefinitely large population and a capital plant growing at high rates. But they can provide the energy base for the sustainable society of the future. They are abundant, widespread, and varied. Their associated pollution flows are lower and generally less harmful than those of fossil or nuclear energy.

If the most sustainable, least polluting sources were developed and used with high efficiency, they could power the needs of the human race without going beyond the limits. It simply requires political will, some technological advances, and modest social change.

Since the (undiscovered) gas reserves appear relatively ample, it seems—at the turn of the millennium—that the most limiting constraints on energy use are on the sink side. The issue of climate change caused by carbon dioxide emissions from energy use is discussed later in the chapter.

Materials

Extracting or harvesting primary natural resources often requires moving or processing large quantities of materials that can modify or damage the environment even though they have no economic value. For example, to get access to metal deposits, mineral ores, or seams of coal . . . requires moving huge amounts of covering materials or overburden. Often crude ores must be processed or concentrated before they become commercial commodities, leaving large amounts of process wastes to be disposed. . . . All such flows are part of a country's economic activity, but most never enter the monetary economy. . . . Economic accounts do not usually include them. The resulting statistics understate the natural resource dependence of an industrial economy.

—WORLD RESOURCES INSTITUTE, 1997

Only 8 percent of the world's people own a car. Hundreds of millions of people live in inadequate houses or have no shelter at all, much less refrigerators or television sets. If there are going to be more people in the world, and if they are to have more or better housing, health services, education, cars, refrigerators, televisions, they will need steel, concrete, copper, aluminum, plastic, and many other materials.

The stream of materials from the earth through the economy and back to the earth can be diagrammed in the same way as the flow of fossil fuels, with one exception. Unlike fossil fuels, materials such as metals and glass do

not turn into combustion gases after use. Either they accumulate somewhere as solid waste, or they are reclaimed and recycled, or they are broken
down, pulverized, leached, vaporized, or otherwise dispersed into soils,
waters, or the air.

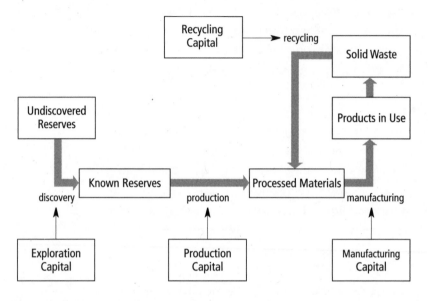

Unknown Reserves to Recycling

Figure 3-16 shows the history of global consumption for five important
metals between the year 1900 and 2000. The consumption data show more
than four-fold growth in use between the years 1950 and 2000.

There is a limit to the amounts of copper, nickel, tin, and related metals
that even wealthy people can use each year. That limit is high, however, at
least if the American lifestyle is indicative. For most metals, the average use
rate of a person in the industrialized world is 8 to 10 times the use rate of
people in the nonindustrialized world. If an eventual nine billion people all
consumed materials at the rate of the average late-twentieth-century
American, that would require an increase in worldwide steel production by a
factor of five, copper by a factor of eight, and aluminum by a factor of nine.

Most people have an intuitive sense that such materials flows are neither
possible nor necessary. They are not possible because of the limits to the
earth's sources and sinks. All along the way from source to sink the processing,

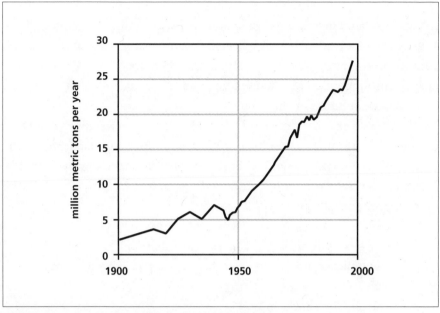

FIGURE 3-16 Global Consumption of Five Important Metals
The consumption of copper, lead, zinc, tin, and nickel grew dramatically during the twentieth century. (Source: Klein Goldewijk and Battjes; U.S. Bureau of Mines; USGS; U.S. CRB.)

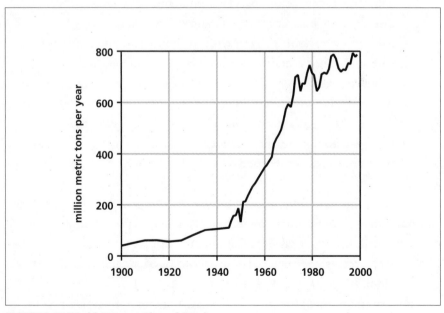

FIGURE 3-17 World Consumption of Steel
The consumption of steel shows S-shaped growth. (Source: Klein Goldewijk and Battjes; U.S. Bureau of Mines; USGS; U.S. CRB.)

fabricating, handling, and use of materials leaves trails of pollution. They are not necessary because the per person materials throughputs of the rich nations of the late twentieth century, like their food, water, wood, and energy throughputs, are wasteful. A good life could be supported with much less destruction of the planet.

There are signs that the world is learning the lesson. Figure 3-17 shows the recent world production history of steel. Something happened in the mid-1970s to interrupt what had been smooth exponential growth trends. There are several theories to explain that reduction in growth rate. All of them appear partially correct.

- The emerging trend toward "dematerialization" was driven by economic incentives and the technological possibility to do more with less.
- The oil price shocks in 1973 and again in 1979 made the prices of energy-intensive metals rise sharply, strengthening the incentives to save on energy and materials in all applications.
- The same higher prices, plus environmental laws and solid waste disposal problems, encouraged materials recycling.
- All of these pressures accelerated a technical revolution. Plastics, ceramics, and other materials were substituted for metals. Products made from metals—automobiles, soft-drink cans, and many others—were made lighter.
- Within the stagnant economy of the 1980s the heavy manufacturing sectors were most depressed, so basic metal demands were reduced disproportionately.[75]

Although the economic reasons for the slower growth in materials consumption may be temporary, the technical changes will probably be permanent, as will the environmental pressures to reduce materials flows. Interestingly, prices for materials have continued to fall during the last decades, indicating that supply outgrew demand.[76]

Poor communities have always reclaimed and reused materials because of scarce sources. Rich communities are relearning how to recycle because of scarce sinks. In the process, recycling is changing from a labor-intensive to a capital- and energy-intensive activity, employing mechanized compost turners, shredders and screening systems, digesters, sludge mixers, reverse vending machines (to return any deposit paid on bottles), and management companies that set up waste recovery programs for industries or municipalities.

Forward-looking manufacturers are designing products from teapots to cars with final disassembly and recycling in mind. A new BMW car, for instance, has a plastic body designed for easy recycling. Plastics are increasingly marked with their resin type, and fewer types are mixed together, so that they can be separated and reused.

Tiny changes multiplied many times can make a big difference. The invention in 1976 of the pop-top opener tab on the aluminum soda can meant that the tab stayed with the can, therefore passing back through the recycling process, rather than being thrown away. Around the turn of the millennium, Americans used some 105 billion (10^9) aluminum cans per year, of which some 55 percent were recycled. That means that *every year* the recycling of those tiny tabs saved 16,000 tons of aluminum and around 200 million kilowatt-hours of electricity.[77]

Separating and recycling materials after use is a step toward sustainability. It begins to move materials through the human economy the way they move through nature—in closed cycles. In nature the waste from one process becomes an input to another process. Whole sectors of ecosystems, particularly in the soils, work to take nature's waste materials apart, separate them into useable pieces, and send them back into living creatures again. The modern human economy is finally developing a recycling sector, too.[78]

But recycling trash is only dealing with the final and least problematic end of the materials stream. A rule of thumb says that every ton of garbage at the consumer end of the stream has also required the production of 5 tons of waste at the manufacturing stage and 20 tons of waste at the site of initial resource extraction (mining, pumping, logging, farming).[79] The best ways to reduce these flows of waste are to increase the useful lifetimes of products and to reduce materials flows at their source.

Increasing product lifetime through better design, repair, and reuse (as, for example, in washing cups instead of using throwaways) is more effective than recycling, because it doesn't require crushing, grinding, melting, purifying, and refabricating recycled materials. Doubling the average lifetime of any product will halve the energy consumption, the waste and pollution, and the ultimate depletion of all the materials used to make it. But drawing final conclusions on what minimizes the ecological footprint requires a thorough life-cycle analysis, which often gives surprising results.

Source reduction means finding a way of performing the same job with less material. It is the equivalent of energy efficiency, and the possibilities are enormous. In 1970 a typical American car weighed more than three tons, nearly all of it metal. Today the average car is much lighter, and much of it is plastic. Computer circuits are carried on minute silicon chips instead of heavy ferromagnetic cores. A small flash drive that fits in your pocket can hold as much information as 200,000 book pages. One hair-thin strand of ultrapure glass can carry as many telephone conversations as hundreds of copper wires and with better sound quality.

Instead of the high temperatures, severe pressures, harsh chemicals, and brute force that have characterized manufacturing processes since the beginning of the industrial revolution, scientists are beginning to understand how to use the intelligence of molecular machines and of genetic programming. Breakthroughs in nanotechnology and biotechnology are beginning to allow industry to carry out chemical reactions the way nature does, by careful fitting of molecule to molecule.

The possibilities for recycling, greater efficiency, increased product lifetime, and source reduction in the world of materials are exciting. On a global scale, however, they have not yet reduced the vast materials flow through the economy. At best, they have slowed its rate of growth. And billions of people still want cars and refrigerators. Though most people currently are more aware of sink limits than of source limits for materials throughputs, continued growth in materials demand will eventually run into source limits as well. Many of the materials most useful to human society occur only rarely in concentrated form in the earth's crust. Their exploitation is posing mounting costs—costs measured in energy, capital, environmental impact, and social disruption.

Geologist Earl Cook illustrated how unusually concentrated, and how rare, most mineable ores are.[80] Technology has improved greatly since Cook completed his analysis almost three decades ago. But the general implications of his study remain valid. Some minerals, such as iron and aluminum, are enormously abundant. They will not be limited by sources, and they can be mined in many areas. Others, like lead, tin, silver, and zinc, are much more limited. For them depletion is a more imminent prospect.

Some impression of relative scarcity is provided by resource and reserve data provided in a recent study of the global mining industry by the International Institute for Environment and Development (IIED). Table 3-2 summarizes data on eight important metals. At 2 percent annual growth (that is high for some materials and low for others—but not a bad average) current reserves would support production for periods ranging from 15 to 80 years. Of course technology will improve and prices will rise, while producers explore new areas and discover new mineable materials. So these reserve life estimates are low. How low? Estimates of crustal abundance suggest productive lifetimes of 500 to 1000 years. The real availability lies somewhere in between. The amount of resources that can be moved into reserves depends on energy and capital cost, as producers are forced to deal with the social and environmental costs of their operations.

TABLE 3-2 Life Expectancies of Identified Reserves for Eight Metals

Metal	Annual production 1997–99 average (million (10^6) metric tons per year)	Annual growth in production 1975–99 average (percent per year)	1999 Identified reserves (billion (10^9) metric tons)	Life expectancy of identified reserves at annual growth in production of 2% per year (years)	Resource base (trillion (10^{12}) metric tons)	Life expectancy of resource base at annual growth in production of 2% per year (years)
Aluminum	124	2.9	25	81	2,000,000	1,070
Copper	12	3.4	0.34	22	1,500	740
Iron	560	0.5	74,000	65	1,400,000	890
Lead	3.1	-0.5	0.064	17	290	610
Nickel	1.1	1.6	0.046	30	2.1	530
Silver	0.016	3.0	0.00028	15	1.8	730
Tin	0.21	-0.5	0.008	28	40.8	760
Zinc	0.8	1.9	0.19	20	2,200	780

This table illustrates the enormous gap between identified reserves and the resource base. Identified reserves are currently known and expected to be mined with available technology and current prices. The resource base is the total amount believed to be present in the earth's crust. Humanity will never be able to exploit the full resource base, but changes in price, technology, and new discoveries will certainly increase the identified reserves. (Source: MMSD.)

The IIED study pointed to the potential role that sinks may come to play in limiting our use of minerals.

> Although trends in minerals production and use and in the esti-mated resource base have reduced concerns that the world is "run-ning out" of minerals, the potential limits that environmental and social factors may place on mineral availability are receiving mounting attention. Developments that may limit the availability of minerals include:
>
> - the availability of energy or the environmental effects of energy use as energy per unit output increases at lower ore grades;
> - the availability of water for minerals production or the environ-mental impacts of using increasing amounts of water at lower ore grades;
> - society's preference to use land for reasons other than mineral pro-duction, whether for biological diversity and pristine wilderness protection, cultural significance, or agriculture and food security;
> - community intolerance of the impacts of the minerals industry;
> - changing patterns of use; and
> - ecosystem limits on the build-up of mineral products or by-products (especially metals) in the air, water, topsoil, or vegetation.[81]

Figure 3-18 shows how the process of mineral depletion proceeds—as in the case of copper's gradually decreasing ore concentration. Figure 3-19 shows the consequence of decreasing ore concentration. As the amount of useable metal in the ore falls, the amount of rock that must be mined, ground up, and treated per ton of product rises with astonishing speed. As the average grade of copper ore mined in Butte, Montana, fell from 30 per-cent to 0.5 percent, the tailings produced per ton of copper rose from 3 tons to 200 tons. This rising curve of waste is closely paralleled by a rising curve of energy required to produce each ton of final material. Metal ore deple-tion hastens the rate of fossil fuel depletion and places greater burdens on the planet's sinks.

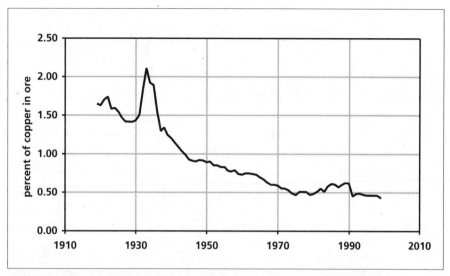

FIGURE 3-18 The Declining Quality of Copper Ore Mined in the United States

Ores averaging between 2 and 2.5 percent copper were mined in the United States before 1910. Since then there has been a persistent decline in average grade. The peak in the 1930s and the slight rise in the 1980s were caused by economic downturns that shut down marginal mines and left functioning only those with the richest ores. (Sources: U.S. Bureau of Mines; USGS.)

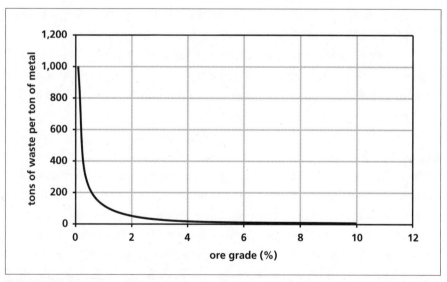

FIGURE 3-19 Depletion of Mineral Ores Greatly Increases the Mining Wastes Generated in Their Production

As the average grade of ore declines through depletion from 8 percent or more to 3 percent, there is a barely perceptible increase in the amount of mining waste generated per ton of final metal. Below 3 percent, waste per ton increases dramatically. Eventually the cost of dealing with the wastes will exceed the value of the metal produced.

Sinks for Pollution and Waste

During the last few decades, humans have emerged as a new force of
nature. We are modifying physical, chemical, and biological systems
in new ways, at faster rates, and over larger spatial scales than ever
recorded on Earth. Humans have unwittingly embarked upon a
grand experiment with our planet. The outcome of this experiment
is unknown, but has profound implications for all of life on Earth.

—JANE LUBCHENCO, 1998

At the time of the 1972 Stockholm Conference on the Environment, there
were no more than 10 nations with environmental ministries or agencies.
Now there are few nations without an environmental bureaucracy. A profu-
sion of environmental educational programs has appeared on the scene,
along with numerous special-interest groups promoting various environmental causes. The record of these relatively new environmental protection
institutions is mixed. It is wrong to conclude that the world has solved its
pollution problems—or that there has been no progress at all.

The greatest successes have come with specific toxins that are unam-
biguously harmful to human health and that can be singled out and simply
banned. Figure 3-20 shows, for instance, that banning the use of lead in U.S.
gasoline has permitted lead concentrations in human blood to decline.
Levels of other pollutants in certain places, such as cesium-137 in Finland
and DDT in Baltic countries, have also fallen over recent decades.

In the industrialized countries, after determined effort and considerable
expense, there has been partial success in decreasing some, but not all, of
the most common air and water pollutants. Figure 3-21 shows that in the G7
nations[82] sulfur dioxide emissions have been cut by almost 40 percent by
scrubbers on smokestacks and shifts to low-sulfur fuels. The pollutants
carbon dioxide and nitrogen oxide are chemically difficult to scrub; they
have been held roughly constant for 20 years, despite economic growth,
mainly because of gains in energy efficiency.

The history of pollutants in the Rhine River provides an excellent illustration of the triumphs and disappointments of water pollution control.

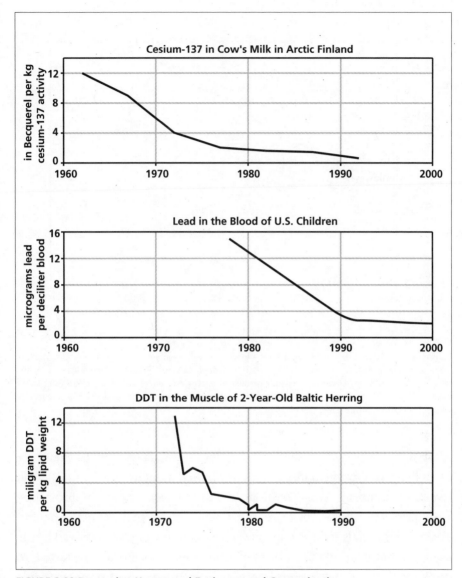

FIGURE 3-20 Decreasing Human and Environmental Contamination
Levels of some pollutants have been dropping over the past few decades in certain places. The most dramatic improvements have come from outright bans on toxic substances such as lead in gasoline and the pesticide DDT, and the halt to nuclear bomb testing in the atmosphere. (Sources: Swedish Environmental Research Institute; AMAP; EPA.)

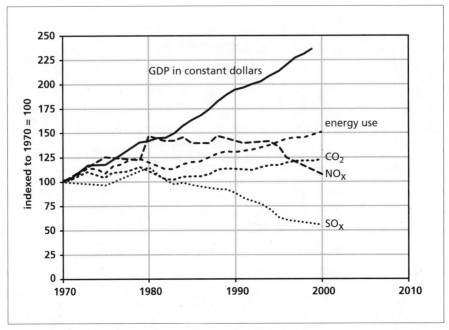

FIGURE 3-21 Trends in Emissions of Selected Air Pollutants
Industrialized countries have made significant efforts to achieve energy efficiency and emission controls. Although their economies (as measured by GDP) have doubled since 1970, their emissions of CO_2 and NO_x have remained almost constant (mainly because of energy efficiency), and their emissions of sulfur oxides (SO_x) have decreased by 40 percent (because of both energy efficiency and active abatement technologies). (Sources: World Bank; OECD; WRI.)

After World War II increasing pollution levels gradually deprived the Rhine of its life-giving oxygen. Oxygen levels reached a minimum around 1970 at levels where no life could be maintained, but were greatly improved by 1980, primarily because of huge investments in sewage treatment systems. Toxic heavy metals such as mercury and cadmium were not removed by the sewage treatment plants, however, and their concentration started to decline only once the nations bordering the Rhine agreed on increasingly strict regulations against pollution. As a result, by 2000 heavy metals had been largely removed from the water. But they still permeate bottom sediments, and since they do not break down chemically, they remain at high levels, particularly in the Rhine delta. Chloride levels have also continued to stay high. The downstream nations did not yet find a way of applying effective pressure against the main chloride source: the salt mines in Alsace—although these may ultimately be closed. Nitrogen pollution from fertilizer drainage off agricultural lands also remains high. Because its sources are too dispersed to be gathered into a sewage treatment system, the only way it can be reduced is by changing farming practices throughout the Rhine watershed. Even so, it was worth celebrating in 1996 when the first salmon reappeared in the upper Rhine valley of Baden-Baden after having been gone for 60 years.[83]

Similarly, other industrialized nations have made major investments to improve the water quality in major rivers and waterways. By investing tens of billion of dollars in waste treatment plants, former cesspools have been upgraded to salmon-carrying quality. The most famous example is probably the Thames River. But even the water in New York Harbor has become cleaner since 1970 (figure 3-22).[84] Cleaner water means, in fact, that emissions per unit of activity were pushed down faster than the rather significant growth in human activity levels. The ecological footprint on the waterways has declined. The same goes for air quality in many industrialized countries. Through a combination of strict regulation, investment in filtering, and change to cleaner production technologies, air pollution levels (for example, of particles, sulfur dioxide, carbon monoxide, and lead) have come down dramatically in the United Kingdom and the United States over the past several decades. Even less tractable pollutants, such as NO_2 and ozone in the lower atmosphere, have been reduced.[85] And once more, this occurred despite rather significant growth in such activities as power generation, heating, and transport of people and goods. There has even been progress in the removal of modern toxins, including PCB, DDT, and other pesticides.[86] But here the success is more local and the total picture mixed, because many of these persistent, bio-accumulative substances are transported across the globe and accumulate in the body fat of distant populations.

That's the record in rich countries with money to spend on pollution abatement. The worst air and water pollution levels in the world are now found in Eastern Europe and the emerging economies, where billion-dollar abatement efforts are simply unimaginable. This fact was drawn to the world's attention in 2001 when a haze darkened the skies of Southeast Asia for weeks.

And that's the record for the most obvious pollutants—the ones people experience directly, and thus the ones that attract political attention. The

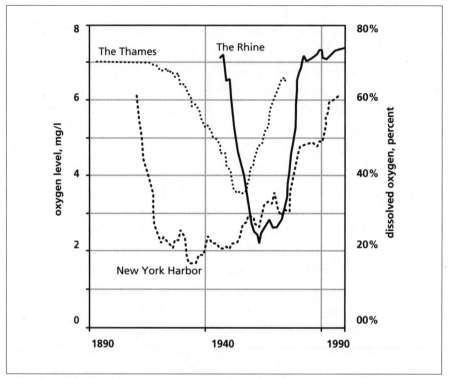

FIGURE 3-22 Oxygen Levels in Polluted Waters
Organic pollution can reduce the level of life-giving oxygen in rivers. Since the 1960s and '70s, large investment in sewage treatment systems have allowed oxygen levels to improve in the Rhine, the Thames, and
in New York harbor. (Source: A. Goudie; P. Kristensen and H. Ole Hansen; OECD; DEP.)

visible water and air pollutants are also increasingly being targeted—rather
successfully—in the current focus on increased eco-efficiency in the
world's environmentally leading corporations. This focus needs to become
permanent, however, in order to counterbalance the continuing increase in
human activity.

The most intractable pollutants, so far at least, are nuclear wastes, hazardous wastes, and wastes that threaten global biogeochemical processes,
such as the greenhouse gases. They are chemically the hardest to sequester
or detoxify, physiologically the hardest for our senses to detect, and economically and politically the most difficult to regulate.

No nation has solved the problem of nuclear wastes. In nature such
wastes are hazardous to all forms of life, both by outright toxicity and muta-
genicity. In the wrong hands they can become instruments of terror. Nature

has no way of rendering them harmless. They disintegrate by their own inner timetable, which can be decades, centuries, or even millennia. As by-products of nuclear power production, they are accumulating steadily, stored underground or in water pools within the containment vessels of nuclear reactors, in the hope that someday the technical and institutional creativity of humankind will come up with someplace to put them. As a consequence, there is widespread, healthy skepticism about the large-scale use of nuclear energy.

Another important class of problem waste is human-synthesized chemicals. They have never before existed on the planet, and therefore no organisms have evolved in nature to break them down and render them harmless. Over 65,000 industrial chemicals are now in regular commercial use. Toxicology data are available on a few of them. Every day new chemicals enter the marketplace, and many of them are not tested thoroughly for toxicity.[87] Every day thousands of tons of hazardous wastes are generated in the world, much of them in the industrialized countries. Slowly there is recognition of the problem; many of these nations have begun efforts to rehabilitate soils and groundwaters poisoned by decades of irresponsible chemical dumping.

Then there are the contaminants that pollute the earth as a whole. These global pollutants, no matter who generates them, affect everyone. A dramatic example has been the effect of the industrial chemicals called chlorofluorocarbons on the stratospheric ozone layer. The ozone story is a fascinating one, because it illustrates humankind's first unambiguous confrontation with a global limit. We think it is so important, and so hopeful, that we tell it fully in chapter 5.

Most scientists, and now many economists as well, believe the next global limit humanity will have to deal with is the greenhouse effect, or global climate change.

> The Earth's climate system has changed, globally and regionally, with some of these changes being attributable to human activities.

- The Earth has warmed 0.6 ± 0.2 degrees Centigrade since 1860 with the last two decades being the warmest of the last century;

- The increase in surface temperatures over the twentieth century for the northern hemisphere is likely to be greater than that for any other century in the last 1000 years;
- Precipitation patterns have changed with an increase in heavy precipitation events in some regions;
- Sea level has risen 10–20 cm since 1900; most non-polar glaciers are retreating; and the extent and thickness of Arctic sea ice is decreasing in summer;
- Human activities are increasing the atmospheric concentrations of greenhouse gases that warm the atmosphere and, in some regions, sulfate aerosols that cool the atmosphere; and
- Most of the observed warming of the last 50 years is attributable to human activities.[88]

For decades scientists have been measuring the accumulation of carbon dioxide in the atmosphere from burning fossil fuels. We published a summary of the CO_2 data already in our first book.[89] It has been known for more than 100 years that carbon dioxide traps heat and increases the temperature of the earth, like a greenhouse that lets the sun's energy in but hinders it from going out. And over the past 30 years it has become ever more obvious that other greenhouse gases emitted by human activity are also building up exponentially in the atmosphere: methane, nitrous oxide, and the same chlorofluorocarbons that are threatening the ozone layer (figure 3-23).

Global climate change is not easy to detect quickly, because the weather from day to day or year to year is naturally variable. Climate is the long-term average of weather; therefore, it can be measured only over the long term. Evidence for global warming, however, was already discernible a decade ago, and has since accumulated at an alarming pace. It is becoming commonplace to read that the last year was the hottest on record—which is not surprising considering the rate of increase in the average global temperature, as illustrated in figure 3-24.

Satellites show a shrinking ice and snow cover over the Northern Hemisphere, the Arctic ice pack is thinning, and Western tourists cruising in a Russian icebreaker were recently surprised to find open water when they arrived at the North Pole. One hundred episodes of "coral bleaching,"

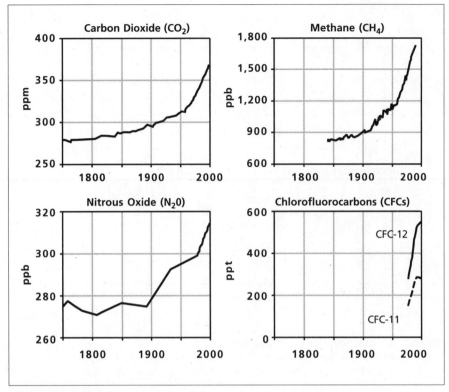

FIGURE 3-23 Global Greenhouse Gas Concentrations
Carbon dioxide, methane, nitrous oxide, and chlorofluorocarbons all reduce emissions of heat from the earth to outer space, thus increasing the temperature of the earth. The atmospheric concentration of these gases—except for CFCs, which were first synthesized in the mid-1900s—has been increasing since the 1800s. (Sources: CDIAC; UNEP.)

where coral reefs around the world turn white and die, were reported during the period 1980 to 1998, compared with only three during the preceding 100 years. Bleaching is a coral reaction triggered quickly by unusual increases in the ocean temperature.[90]

Even some economists—a group well known for its skepticism about "environmentalist alarmism"—are becoming convinced that something unusual and significant is going on in the atmosphere, and that it may have human causes. In 1997 a group of at least 2000 economists, including 6 Nobel laureates issued a declaration:

The balance of evidence suggests a discernible human influence on global climate. As economists, we believe that global climate change

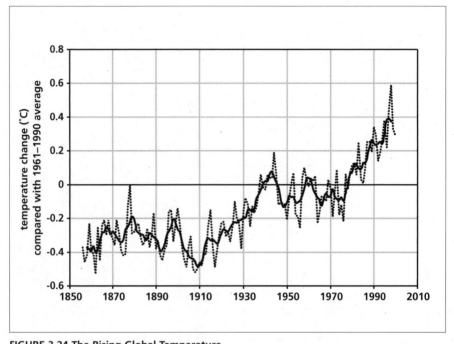

FIGURE 3-24 The Rising Global Temperature
The global average temperature has risen over the past century by some 0.6°C. The dashed line represents annual averages; the thick line represents five-year running averages. (Source: CDIAC.)

carries with it significant environmental, economic, social, and geopolitical risks, and that preventive steps are justified.[91]

One reason for the increasing concern of economists may be that it is possible to observe a disturbing upward trend in the measurable economic losses from weather-related disasters, starting from around 1985 (figure 3-25).

None of the observations above *proves* that the ongoing climate change has human causes. Even if it has, the effects of global climate change on future human activity or ecosystem health cannot be predicted with certainty. Some have exploited that uncertainty in an effort to create a state of confusion,[92] and thus it is important to state clearly what we do know. In this we rely on the several hundred scientists and researchers who make up the UN Intergovernmental Panel on Climate Change, which issues their carefully considered views approximately every five years:[93]

- It is certain that human activities, especially fossil fuel burning

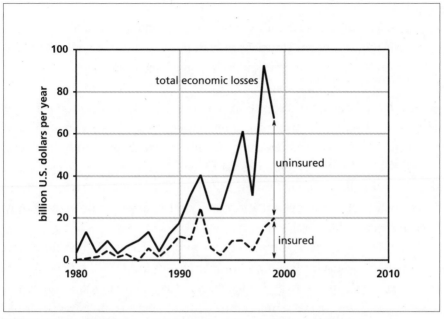

FIGURE 3-25 Worldwide Economic Losses from Weather-Related Disasters
The last two decades of the twentieth century were marked by increasing economic losses from weather-related disasters. (Source: Worldwatch Institute.)

and deforestation, contribute to the atmospheric concentration of greenhouse gases.

- It is certain that the concentration of carbon dioxide (the prime greenhouse gas) in the atmosphere is increasing exponentially. The CO_2 concentration has been monitored for decades. Its historical concentration can be measured from bubbles of air caught in layers of ice drilled from the polar icecaps.
- Greenhouse gases trap heat that otherwise would escape from the Earth into space. That is a well-known property of their molecular structure and spectroscopic absorption frequencies.
- Trapped heat will increase the temperature of the Earth over what it would otherwise be.
- The warming will be unequally distributed, more near the poles than near the equator. Because the Earth's weather and climate are largely driven by temperature differences between the poles and the equator, winds, rains, and ocean currents will shift in strength and direction.

- On a warmer Earth the ocean will expand and sea levels will rise. If the warming is sufficient to melt polar ice in large quantities, sea levels will rise significantly, but on a longer time horizon.

There are three large uncertainties. One is what the global temperature would have been without human interference. If long-term climatological factors unrelated to the increase in greenhouse gases happen to be warming the planet, then the greenhouse gases will strengthen those factors. A second uncertainty is what, exactly, a warming planet will mean for temperatures, winds, currents, precipitation, ecosystems, and the human economy in each specific place on Earth.

The third large uncertainty has to do with feedbacks. Carbon flows and energy flows on planet Earth are immensely complex. There may be self-corrective mechanisms, negative feedback processes, that will stabilize the greenhouse gases or the temperature. One of them is already operating: The oceans are absorbing about half the excess carbon dioxide emitted by humanity. That effect is not strong enough to stop the rise in the atmospheric carbon dioxide concentration, but it is sufficient to slow it.

There may also be destabilizing positive feedback loops, which, as the temperature rises, will make things even warmer. For example, as warming decreases the snow and ice cover, the Earth will reflect away less heat from the sun, thereby warming further. Melting tundra soils could release huge amounts of frozen methane, a greenhouse gas, which will cause more warming, more melting, and the release of still more methane.

No one understands how the many possible negative and positive feedback responses to the rise in greenhouse gases will interact or whether the positive or the negative will dominate. Luckily, the 1990s saw a huge increase in the scientific exploration of these issues, and computer simulations are making ever better forecasts of likely climate effects.[94] The resulting "weather forecasts for 2050" are sufficiently worrying to grab the attention of the public.

> The question is not whether climate will change further in the future in response to human activities, but rather by *how much* (magnitude), *where* (regional patterns), and *when* (the rate of change). It

is also clear that climate change will, in many parts of the world, adversely effect socio-economic sectors, including water resources, agriculture, forestry, fisheries and human settlements, ecological systems (particularly coral reefs), and human health (particularly vector-borne diseases). Indeed, the IPCC Third Assessment Report concluded that most people will be adversely affected by climate change.[95]

Scientists do know that there have been temperature upheavals on earth in the past, and that they have not been quickly self-correcting or smooth or orderly. In fact they have been chaotic. Figure 3-26 shows a 160,000-year history of the earth's temperature and of the atmospheric concentrations of two greenhouse gases, carbon dioxide and methane.[96] Temperature and greenhouse gases have varied together, though it's not clear which causes which. Most probably each causes the other in a complicated set of feedback loops.

But the most important message in figure 3-26 is that *current* atmospheric concentrations of carbon dioxide and methane are *far higher than they have been for 160,000 years.* Whatever the consequences might be, there is no question that humanity's emissions of greenhouse gases are suddenly filling up the atmospheric sinks much faster than the planet can empty them. There is a significant disequilibrium in the global atmosphere, and it is getting exponentially worse.

The processes set in motion by this disequilibrium may move slowly, as measured by human time scales. It may take decades for the consequences to be revealed in melting ice, rising seas, changing currents, shifting rainfall, greater storms, and migrating insects or birds or mammals. But it is also plausible that climate may change suddenly, through positive loops that we do not yet understand. In 2002 a National Academy of Sciences committee reported:

> Recent scientific evidence shows that major and widespread climate changes have occurred with startling speed. For example, roughly half the north Atlantic warming since the last ice age was achieved in only a decade, and it was accompanied by significant climatic changes across most of the globe. . . . The abrupt changes of the past are not fully explained yet.[97]

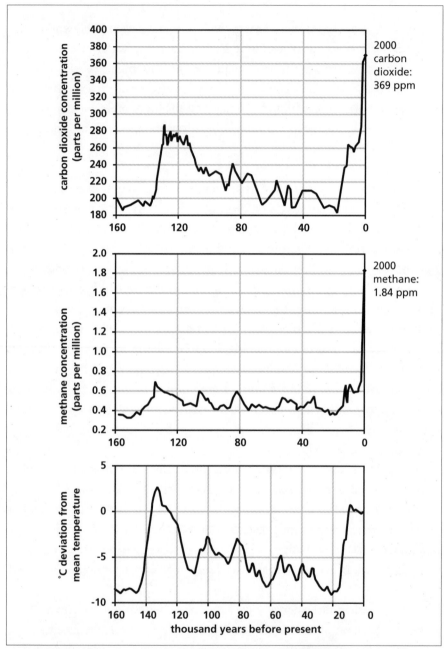

FIGURE 3-26 Greenhouse Gases and Global Temperature Over the Past 160,000 Years
Ice core measurements show that there have been significant temperature variations on Earth (ice ages and interglacial periods), and the carbon dioxide and methane levels in the atmosphere have varied in concert with global temperature. Recent concentrations of these greenhouse gases have soared much higher than they have been since long before the appearance of the human species. (Source: CDIAC.)

Whether the onset is slow or fast, we know that it will take centuries, perhaps millennia, for the negative consequences to be reversed.

The negative environmental impacts of human activity that we have discussed in this chapter were not necessary. They were all avoidable. Increasingly, pollution is no longer seen as a sign of progress, but as a sign of inefficiency and carelessness. As industries realize that, they are quickly finding ways to reduce their emissions and resource use by rethinking manufacturing processes from beginning to end, moving from "end-of-pipe solutions" (reducing the emissions from ongoing production processes) to "cleaner production" (designing the product and production processes in order to minimize emissions and resource use), to "industrial ecology" (using the effluents from one factory as raw material for another). A circuit-board manufacturer invests in ion-exchange columns to reclaim heavy-metal wastes and ends up with an income from the recycled metals, a much-reduced water bill, and lower liability insurance. A manufacturing company reduces its air pollution emissions, its water pollution emissions, its water requirements, and its solid waste production, and saves hundreds of million a year in operating expenses. A chemical firm decides to reduce its CO_2 emissions to avoid anticipated emission fees and makes dramatic savings in energy costs at the same time.

Much of this work, perhaps somewhat surprisingly, has proven to be profitable, even in the short run, over and above the positive public relations that go with such changes. The economic benefits will, no doubt, provide a strong argument for the continued reduction of the ecological footprint per unit of consumption.

If the average lifetime of each product flowing through the human economy could be doubled, if twice as many materials could be recycled, if half as much material needed to be mobilized to make each product in the first place, that would reduce the throughput of materials by a factor of eight.[98] If energy use became more efficient, if renewable energy sources were used, if land, wood, food, and water were used less wastefully and forests were restored, that would stop the rise of greenhouse gases and of many other pollutants.

Beyond the Limits

A rough assessment . . . shows that current appropriations of natural resources and services already exceed Earth's long-term carrying capacity. . . . If everybody on Earth enjoyed the same ecological standards as North Americans, we would require three earths to satisfy aggregate material demand, using prevailing technology. . . . To accommodate sustainably the anticipated increase in population and economic output of the next four decades, we would need six to twelve additional planets.

—MATHIS WACKERNAGEL AND WILLIAM REES, 1996

The evidence we have given in this chapter, plus much more contained in the world's databases, plus daily reports in the media all show that the human economy is not using the earth's stocks and sinks sustainably. Soils, forests, surface waters, groundwaters, wetlands, atmosphere, and the diversity of nature are being degraded. Even in places where renewable resource stocks appear to be stable, such as the forests of North America or the soils of Europe, the quality, diversity, or health of the stock is in question. Pollutants are accumulating; their sinks are overflowing. The chemical composition of the entire global atmosphere is being changed in ways that are already measurably perturbing the climate.

Living on Capital, Not Income

If only one or a few resource stocks were falling while others were stable or rising, one might argue that traditional growth could continue by the substitution of one resource for another (though there are limits to such substitution). If only a few sinks were filling, humanity might substitute one (say, the ocean) for another (say, the air). But since many sinks are filling and many stocks are declining, and the human ecological footprint has surpassed the sustainable level, we need a more fundamental change.

The limits that exist, let us be clear, are not limits to the level of human economic activity as measured by the gross world product. The limits that exist constrain the ecological footprint of human activity. And these limits are not absolute in the short run. To go beyond the limits does not mean running into an absolute wall. The simplest analogy is that of ordinary fisheries, where the annual catch can exceed the annual regrowth for quite a while— actually until the stock of fish has been decimated. Similarly, emissions of greenhouse gases can go on growing for a while, even when sustainability limits are exceeded, before negative feedbacks from climate change force emissions down. But down is the direction that throughputs will eventually have to go after the overshoot, by human choice or by natural limits.

Many people recognize on a local level that the human footprint has grown beyond local limits. Jakarta emits more air pollution than human lungs can bear. The forests in the Philippines are nearly gone. The soils of Haiti have been worn down in places to bare rock. The cod fisheries off Newfoundland have been closed. Parisians have to endure summer days of reduced speed limits to cut down pollution from their fuming cars. Several European countries saw thousands die prematurely as the summer of 2003 set new records for high temperatures. The chemical load in the Rhine was for many years so high that dredged silt from Dutch harbors now has to be treated as hazardous waste. Skiers visiting Oslo in the winter of 2001 found hardly any useful snow.

In the case of particular problems, such as the CFCs that erode the ozone layer, there has been not only recognition of an overshoot, but determined international effort to take corrective action. And the global effort to limit greenhouse gas emissions is trudging along, albeit steadily hampered by selfish and shortsighted governments representing their equally shortsighted and selfish donors. The Kyoto process certainly illustrates the challenge in moving back from overshoot.

Still there is little discussion about the *general problem* of overshoot, little pressure for the technical changes urgently necessary to make throughputs more efficient, and almost no willingness to deal with the driving forces of population and capital growth. The lack of attention to overshoot could perhaps be excused in 1987. Then even informed groups such as the World Commission on Environment and Development, which looked hard at world trends and labeled them "simply unsustainable," did not find it politically opportune to say *The human world is beyond its limits,* much less grapple seriously with the question of what to do. Possibly because they could not believe it was true. But now, at the turn of the millennium, it is inexcusable to deny the awful reality of overshoot and ignore the consequences.

The reasons for avoiding the issue of overshoot are understandable and political. Any talk of reducing growth feeds into a bitter argument about distribution—of available resources and of responsibility for the current state of affairs. Generally speaking, the ecological footprint of a rich person is much greater than the ecological footprint of a poor person. One German, the saying goes, has a footprint 10 times that of a Mozambiquean, while one Russian draws as many resources from the planet as one German—without even getting a decent standard of living out of it. If the world as a whole is exceeding its limits, who should do something about it: the wasteful rich or the multiplying poor or inefficient ex-socialists? As far as the planet is concerned, the answer is all of the above.

> The continued poverty of the majority of the planet's inhabitants and excessive consumption by the minority are the two major causes of environmental degradation. The present course is unsustainable and postponing action is no longer an option.[99]

Environmentalists sometimes summarize the causes of environmental deterioration with a formula they call IPAT:

$$\text{Impact} = \text{Population} \times \text{Affluence} \times \text{Technology}$$

The impact (I) of any population or nation upon the planet's sources and sinks is related to the product of its population (P) times its level of affluence (A) times the damage done by the particular technologies (T) chosen to support that affluence. In order to reduce the ecological footprint of humanity, it would seem reasonable that every society should make improvements where it has the most opportunity to do so. The South has the most room for improvement in P, the West has the most room for improvement in A, while Eastern Europe has the most room for improvement in T.

The total scope for improvement is astonishing. If we define each term in the IPAT equation more precisely, we can see how many ways there are to reduce the ecological footprint, and what great reductions are possible (see table 3-3).[100]

Affluence is determined by a high rate of consumption; for example, the number of hours spent watching television, driving a car, or relaxing in a room. The ecological footprint of affluence is the impact or *throughput* generated by the material, energy, and emissions associated with this consumption. For instance, if one drinks three servings of coffee per day, the footprint can be widely different depending on whether traditional china or plastic cups are used. Maintaining the china cups takes water and soap to wash them and a small flow of cups to replace annual breakage. If a person uses and discards polystyrene cups, on the other hand, the maintenance flow includes all the cups used in a year as well as the petroleum and chemicals needed to make the polystyrene and transport the cups to their point of use.

The impact of technology is defined in table 3-3 as the energy needed to make and deliver each material flow, multiplied by the environmental impact per unit of energy. It takes energy to mine the clay for ceramic cups, to fire the clay, to deliver the cups to the household, and to heat the water to wash them. It takes energy to find and pump the oil for polystyrene cups, to transport the oil, run the refinery, form the polymer, mold the cups, deliver the cups, and transport the used cups to the dump. Each kind of energy has its environmental impact. The ecological footprint can be technologically changed with pollution control devices, with energy efficiency changes, or by switching to another energy source.

Changes in any factor in table 3-3 will change the ecological footprint and bring the human economy closer to or farther away from the earth's limits. Reducing population or the stock of material accumulated by each person will help keep the human world within the limits of the planet. So will higher eco-efficiency, which is lower rates of energy or material—and lower emissions—per unit of consumption. The table lists some of the tools that might help reduce each factor in the equation and also shows some guesses about how much each factor contributing an impact might be reduced, and over what time period.

TABLE 3-3 The Environmental Impact of Population, Affluence, and Technology

$$\text{population} \times \frac{\text{capital stock}}{\text{person}} \times \frac{\text{material throughput}}{\text{capital stock}} \times \frac{\text{energy}}{\text{material throughput}} \times \frac{\text{environmental impact}}{\text{energy}}$$

Example:

$$\text{population} \times \frac{\text{cups}}{\text{person}} \times \frac{\text{water} + \text{soap}}{\text{cups/year}} \times \frac{\text{gigajoules or kilowatt-hours}}{\text{kilogram water} + \text{soap}} \times \frac{CO_2,\ NO_x\ \text{land use}}{\text{gigajoules or kilowatt-hours}}$$

	Population	Affluence		Technology	
		capital stock / person	material throughput / capital stock	energy / material throughput	environmental impact / energy
Applicable Tools	Family planning Female literacy Social welfare Role of women Land tenure	Values Prices Full costing What do we want? What is enough?	Production longevity Material choice Minimum-materials design Recycle, reuse Scrap recovery	End-use efficiency Conversion efficiency Distribution efficiency Systems integration Process redesign	Benign sources Scaling Siting Technical mitigation Offsets
Approximate Scope of Long-Term Change	~2x	?	~3–10 x	~5–10 x	~10^2–10^{3+} x
Time Scale of Major Change	~50–100 years	~0–50 years	~0–20 years	~0–30 years	~0–50 years

Presented this way, it is clear that there are many, many choices. Human impact on the planet's sources and sinks could be reduced to an astounding degree. Even assuming only minimal achievement in each area of possible change, taken together they could reduce the human impact on the planet by a factor of *several hundred or more*.

If there are so many options, why are we not going to much trouble to pursue any of them? What if we did? What would happen if population, affluence, and technology trends began to turn around? What about the ways they are interconnected with each other? What happens if the ecological footprint is reduced by technical change, but then population and capital grow still farther? What happens if the ecological footprint isn't reduced at all?

These are questions not about resource stocks and pollution sinks viewed separately, as we have seen them in this chapter, but about the whole ecological footprint, interacting with population and capital, which are in turn interacting with each other. To address these questions we need to move on from a static, one-factor-at-a-time analysis to a dynamic whole-system analysis.

CHAPTER 4

World3: The Dynamics of Growth in a Finite World

'If current predictions of population growth prove accurate and patterns of human activity on the planet remain unchanged, science
and technology may not be able to prevent either irreversible degradation of the environment or continued poverty for much of the
world.

—Royal Society of London and

U.S. National Academy of Sciences, 1992

The factors responsible for growth in population and industry involve
many long-term trends that reinforce and conflict with each other. Birth
rates are coming down faster than expected, but the population is still rising.
Many people are getting richer; they are demanding more industrial products. But they also want less pollution. The flows of energy and materials
required to sustain industrial growth are depleting nonrenewable resource
stocks and deteriorating renewable resources. But there is steady progress
in developing technologies that discover new reserves and use materials
more efficiently. Every society confronts a shortage in capital; investments
are needed to find more resources, produce more energy, clean up pollution, improve schools, health care, and other social services. But those
investments must compete with an ever growing demand for more consumer goods.

How will these trends interact and evolve over the coming decades? To
understand their implications, we need a model much more complex than
the ones in our heads. This chapter 4 is about World3, the computer model
we have created and used. We summarize here the main features of
World3's structure and describe several important insights it gives us about
the twenty-first century.

The Purpose and Structure of World3

The universal desire for certainty about what is to come can lead to misunderstanding and frustration when someone presents a model as the basis for talking about the future. We have had that trouble since we published the first edition of this book, more than 30 years ago. The problem is illustrated in a classic science fiction novel through a conversation between a modeler, named Seldon, and his emperor.

> "I am given to understand that you believe it possible to predict the future."
>
> Seldon suddenly felt weary. It seemed as though this misinterpretation of his theory was constantly going to occur. Perhaps he should not have presented his paper.
>
> He said, "Not quite, actually. What I have done is much more limited than that. . . . What I have done . . . is to show that . . . it is possible to choose a starting point and to make appropriate assumptions that will suppress the chaos. That will make it possible to predict the future, not in full detail, of course, but in broad sweeps; not with certainty . . ."
>
> The Emperor, who had listened carefully said, "But doesn't that mean that you have shown how to predict the future?"[1]

In the remainder of this book we will often use World3 to generate scenarios that help us talk about the "broad sweeps" of the future. To minimize confusion about our goals, we start with several definitions and cautionary notes about models.

A model is a simplified representation of reality. If it were a perfect replica, it would not be useful. For example, a road map would be of no use to drivers if it contained every feature of the landscape it represents—it focuses on roads and omits, for example, most features of buildings and plants along the way. A small physical airplane model can be useful for exploring the dynamics of a particular airfoil in a wind tunnel, but it gives no information about the comfort of passengers in the eventual operational plane. A painting is a graphic model that may convey a mood or the physical placement of features on a landscape. But it does not answer any questions about the cost or the insulation of the buildings it portrays. To deal with those issues a different graphic model would be required—an architect's construction blue print. Because models are always simplifications, they are never perfectly valid; no model is completely true.

Instead the goal is to create a model that is useful for some specific purpose, for answering a specific set of interrelated questions. Then one must keep in mind the limitations of the model and be aware of all the questions it does not answer. We have focused our efforts on making World3 useful—for a carefully bounded set of questions about long-term physical growth on the planet. Unfortunately, that means World3 will not provide useful answers to most of the questions that concern you.

Models take many forms—common forms are mental, verbal, graphical, mathematical, or physical. For example, many words in this book are verbal models. *Growth, population, forest,* and *water* are just symbols, simple verbal representations that stand for very complex realities. Every graph, chart, map, and photograph is a graphical model. Its relationships are expressed through the appearance and location of objects on the paper. World3 is a mathematical model. The relationships it contains are represented through a set of mathematical equations. We have not used physical models in our effort to understand growth and limits, though they are useful for many other purposes, such as in designing communities or industrial products.

Mental models are the abstractions carried in minds. They are not directly accessible by others; they are informal. Formal models exist in a form that can be directly viewed, and sometimes manipulated, by others. The two should ideally interact. Using formal models, we can learn more about reality and about others' mental models. And that enriches our own mental models. As we learn, we are able to create more useful formal models. That process of iteration has engaged us for more than 30 years. And this book is one result.

To create this book, we have assembled words, data, graphs, and computer scenarios. The book is a model of what is in our minds, and creating it has altered what we know. This text is our best attempt to symbolize our current thoughts and understanding about physical growth on this planet over the coming century. But this book is only a model of those thoughts, which are themselves, like every person's thoughts, only models of the "real world."

Therefore we have a difficulty. We will talk about a formal model, a computer-based simulation of the world. For this model to be of any use, we will have to compare it to the "real world," but neither we nor you, our readers, have one agreed-upon "real world" to compare it to. All of us have only our mental models of the entity that is normally called the real world. Mental models of the surrounding world are informed by objective evidence and subjective experience. They have allowed *Homo sapiens* to be a tremendously successful species. They have also gotten people into many kinds of trouble. But whatever their strengths and weaknesses, human mental models are ludicrously simple compared with the immense, complex, ever-changing universe they try to represent.

To remind ourselves, and you, of our inevitable dependence upon models, we will put the World3 model's referent, the "real world," in quotation marks. What we mean by *"real world"* or *"reality"* is just the shared mental model of the authors of this book. The word *reality* can never mean anything more than the mental model of the user of that word. We can't escape that fact. We can only claim that through the exercise of working with our computer model, our mental models have been forced to become more rigorous, more comprehensive, and more clear than they were before. That's the advantage of computer models: They force some discipline, logic, and basic accounting that is hard to achieve with mental models alone. And they give a much more useful basis for improving mental models.

World3 is complex, but its basic structure is not difficult to understand. It keeps track of stocks such as population, industrial capital, persistent pollution, and cultivated land. In the model those stocks change through flows such as births and deaths (in the case of population); investment and depreciation (in the case of each capital stock); pollution generation and pollution assimilation (in the case of persistent pollution); and (in the case of arable land) land erosion, land development, and land removed for urban and industrial uses. Only a fraction of arable land is cultivated. Multiplying the amount of land cultivated by the average land yield gives total food production. Food production divided by population gives food per capita. If food per capita falls below a critical threshold, the death rate begins to go up.

The components and relationships in World3 are straightforward when viewed one at a time. For example, World3 takes into account the momentum of population growth, the accumulation of pollution, the long lifetime of capital plants, the competition for investment among different sectors. It focuses especially on the time it takes for things to happen, the delays in flows, and the slow unfolding of physical processes. It comprises many, many dozens of feedback loops. These loops are closed chains of causality within which an element is often the partial cause of its own future behavior. A change in population, for example, may cause a change in the economy. As economic output changes composition, it will affect birth and death rates. Those rates will change population even further. The feedback loops are one feature that make World3 dynamically complex.

Another feature are its many *nonlinear* relationships. Such relationships cannot be drawn with straight lines; they do not produce proportional changes over all ranges of related variables. Suppose that A influences B. In a linear relationship, if doubling A causes B to double, then you know that halving A will reduce B by 50 percent. Increasing A by five times will quintuple B. Linear relationships tend to produce behavior that is relatively easy to understand. But linearity is seldom found in the "real world." For example, in World3 we must represent the influence of food per capita on human life expectancy. One relationship between the two is shown in figure 4-1. If people who are inadequately nourished get more food, their life expectancy can increase greatly. Societies that have managed to double average daily food consumption from 2,000 to 4,000 vegetable calorie equivalents per person per day may see their average life expectancy increase by 50 percent—rising from 40 to 60 years. But doubling consumption again, to 8,000, is associated with rather little gain in life expectancy—perhaps an additional 10 years. At some point further gains in food consumption may actually decrease life expectancy.

Nonlinear relationships such as this are found throughout the "real world" and therefore throughout World3. An example of a nonlinear relationship used in World3 is shown in figure 4-2: the cost of developing new agricultural land, as a function of the potentially arable land remaining unused. We assume that the first farmers moved onto the most fertile and well-watered plains and started planting with little cost. This is shown at the extreme right edge of the curve, where almost 100 percent of the potentially arable land still remains undeveloped. But as more land is developed

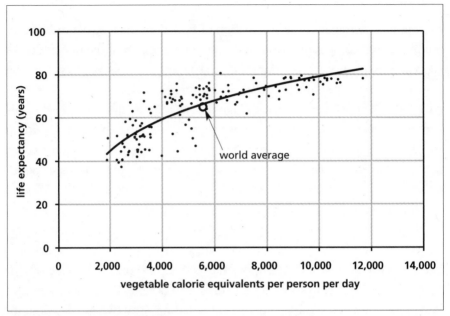

FIGURE 4-1 Nutrition and Life Expectancy
The life expectancy of a population is a nonlinear function of the nutrition that a population receives. Each point on this graph represents the average life expectancy and nutritional level of one nation in 1999. Nutritional level is expressed in vegetable calorie equivalents per person per day; calories obtained from animal sources are multiplied by a conversion factor of seven (since about seven calories of vegetable feed are required to produce one calorie of animal origin). (Sources: FAO; UN.)

for farming (moving toward the left on the graph), what remains is drier or steeper, or has thinner soils or less favorable temperatures. The cost of coping with these problems raises the development cost of the land. In consonance with the classic economic principle that consumers take the lowest-cost goods first, World3 assumes that the last lands to be brought under the plow will cost a great deal indeed—the cost soars nonlinearly.

One thing pushes on another and produces an effect. It pushes a little harder and instead of producing a proportionately larger effect, there's no change, or there's a much bigger change, or a change in the opposite direction. Because of these nonlinearities, both the "real world" and World3 can sometimes produce surprising behavior, as we'll demonstrate later in this chapter.

World3's delays and nonlinearities and feedback loops make it dynamically complex, but the model is still a great simplification of reality. It does not distinguish among different geographic parts of the world, nor does it represent separately the rich and the poor. Pollution is highly simplified in the model. Production processes emit many thousands of different pollu-

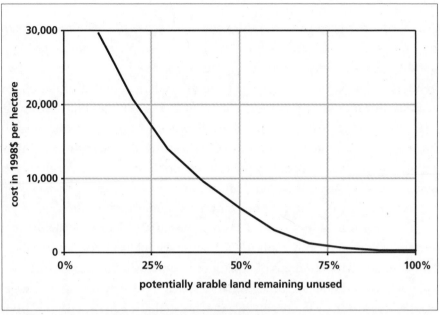

FIGURE 4-2 Development Costs of New Agricultural Land
World3 assumes that the cost of bringing new land into agricultural use increases as the amount of potentially arable land drops. (Source: D. L. Meadows et al.)

tants. These move through the environment at different speeds, affecting plant and animal species in many different ways. World3 captures the influence of these pollutants with two aggregate variables—one representing short lived air pollution and the other long lived toxic materials. The model distinguishes the renewable sources that produce food and fiber from the nonrenewable ones that produce fossil fuels and minerals, but it doesn't keep separate account of each type of food, each fuel, each mineral. It omits the causes and consequences of violence. And there is no military capital or corruption explicitly represented in World3.

That degree of simplicity surprises some people who assume that a world model ought to contain everything we know about the world, especially all the distinctions that are so fascinating and, from the point of view of each academic discipline, so critical. Incorporating those many distinctions, however, would not necessarily make the model better. And it would make it very much harder to comprehend. Despite its relative simplicity World3 is very much more comprehensive and complex than most models that are used to make statements about the globe's long term future.

If you are trying to understand the future behavior of a social system, you need a balanced model. It does not make sense to create models with enormous detail in one sector but grossly simplified assumptions in another. For example, some demographic models keep track of the two genders and account for many age cohorts in a large number of nations or regions. But they simply assume that birth- and death rates will follow pre-determined paths independent of each other.[2] Some economic models include dozens or even hundreds of sectors of the economy, but they assume simple linear relationships between inputs and outputs; or that markets quickly equate supply and demand; or that people make decisions on the basis of pure economic optimization and perfect information.

If a model is to give useful insights about the future behavior of a system, it should portray explicitly the causes of all its important variables. Some models have hundreds of equations to portray influences on one variable or sector, then leave other variables, such as energy use, as an exogenous factor driven by factors outside the model that are in turn derived from historical data or the modelers' intuition. Models, like metal chains, may be limited by their weakest link. We have worked to make the various sectors of World3 equally strong. We have done our best to avoid making simplistic assumptions, leaving out crucial factors, or making important variables dependent on exogenous inputs.

It is not necessary to take our word for this. We have created a World3 CD-ROM disk with the model and documentation. You may obtain a copy of the disk, reproduce all our scenarios, compare them, and evaluate our interpretations of what they mean.[3]

The Purpose of World3

To avoid creating impenetrable thickets of assumptions, modelers must discipline themselves. They cannot put into their models all they know; they have to put in only what is *relevant for the purpose of the model*. The art of modeling, like the arts of poetry or architecture or engineering or map-making, is to include just what is necessary to achieve the purpose, and no more. That is easy to say and hard to do.

Therefore to understand a model and judge its utility, it's important to understand its purpose. We developed World3 to understand the broad sweep of the future—the possible modes, or behavior patterns, through which the human economy will interact with the carrying capacity of the planet over the coming century.[4] Of course, there are many other important long-term global questions to ask: What policies might maximize the industrial development possibilities for Africa? What is the best design for a family planning program in a region where many people are illiterate? How can society close the gap between the rich and the poor within and between nations? Will conflict or negotiation become the dominant means for resolving disputes among nations? The factors and relationships needed to answer those questions are largely missing from World3. Other models, including other computer models, might help answer some of those questions. But if they are to be useful, those models must take into account the answers we generate to World3's core question: *How may the expanding global population and material economy interact with and adapt to the earth's limited carrying capacity over the coming decades?*

To be more specific, the carrying capacity is a limit. Any population that grows past its carrying capacity, overshooting the limit, will not long sustain itself. And while any population is above the carrying capacity, it will deteriorate the support capacity of the system it depends upon. If regeneration of the environment is possible, the deterioration will be temporary. If regeneration is not possible, or if it takes place only over centuries, the deterioration will be effectively permanent.

A growing society can approach its carrying capacity in four generic ways (see figure 4-3).[5] First, it can grow without interruption, as long as its limits are far away or are growing faster than the population. Second, it can level off smoothly below the carrying capacity, in a behavior that ecologists call logistic, or S-shaped, or sigmoid, growth, shown in figure 4-3b. Neither of those options is any longer available to the global society, because it is already above its sustainable limits.

The third possibility for a growing society is to overshoot its carrying capacity without doing massive and permanent damage. In that case the ecological footprint would oscillate around the limit before leveling off. This behavior, illustrated in figure 4-3c, is called damped oscillation. The fourth possibility is to overshoot the limits, with severe and permanent damage to

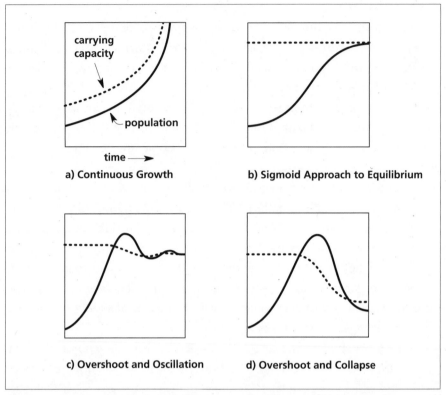

FIGURE 4-3 Possible Modes of Approach of a Population to Its Carrying Capacity
The central question addressed by the World3 model is: Which of these behavior modes is likely to be the result as the human population and economy approach the global carrying capacity?

the resource base. If that were to occur, the population and the economy would be forced to decline rapidly to achieve a new balance with the recently reduced carrying capacity at a much lower level. We use the phrase *overshoot and collapse* to designate this option, shown in figure 4-3d.

There is pervasive and convincing evidence that the global society is now above its carrying capacity. What policies will increase the chances of a smooth transition back beneath planetary limits—a transition like 4-3c rather than 4-3d?

Our concept of the "global society" incorporates the effects of both the size of the population and the size and composition of its consumption. To express this concept we use the term *ecological footprint* that has been defined by Mathis Wackernagel and his colleagues.[6] As we have indicated, the ecological footprint of humanity is the total burden humankind places on the earth. It includes the impact of agriculture, mining, fish catch, forest harvest, pollution emissions, land development, and biodiversity reductions. The ecological footprint typically grows when the population grows, because it grows when consumption increases. But it can also shrink when appropriate technologies are utilized to reduce the impact per unit of human activity.

The concerns motivating our development of World3 may be expressed another way. Given that the ecological footprint of the global population is presently above the earth's carrying capacity, will current policies lead us to a relatively peaceful, orderly oscillation, without forcing drastic declines in population and economy? Or will the global society experience collapse? If collapse is more likely, when might it come? What policies could be implemented now to reduce the pace, the magnitude, the social and ecological costs of the decline?

These are questions about broad behavioral possibilities, not precise future conditions. Answering them requires a different kind of model than does precise prediction. For example, if you throw a ball straight up into the air, you know enough to describe what its general behavior will be. It will rise with decreasing speed, then reverse direction and fall faster and faster until it hits the ground. You know it will not continue to rise forever, nor begin to orbit the earth, nor loop three times before landing.

If you wanted to predict exactly how high the ball would rise or precisely where and when it would hit the ground, you would need precise information about many features of the ball, the altitude, the wind, the force of the initial throw, and the laws of physics. Similarly, if we wanted to attempt to predict the exact size of the world population in 2026, or forecast when world oil production will peak, or specify precisely the rate of soil erosion in 2070, we would need a much more complicated model than World3.

To our knowledge no one has come close to making such a model; nor do we believe anyone will ever succeed. It is simply not possible to make accurate "point predictions" about the future of the world's population, capital, and environment several decades from now. No one knows enough to do that, and there are excellent reasons to believe they never will. The global social system is horrendously and wonderfully complex, and many of its crucial parameters remain unmeasured. Some are probably unmeasurable. Human understanding of complex ecological cycles is very limited.

Moreover, the capacity for humans to observe, adapt and learn, to choose, and to change their goals makes the system inherently unpredictable.

Therefore, when we constructed our formal world model, it was not to make point predictions, but rather to understand the broad sweeps, the behavioral tendencies of the system. Our goal is to inform and to influence human choice. To accomplish these goals, we do not need to predict the future precisely. We need only identify policies that will increase the likelihood of sustainable system behavior and decrease the severity of future collapse. A *prediction* of disaster delivered to an intelligent audience with the capacity to act would, ideally, defeat or falsify itself by inducing action to avoid the calamity. For all those reasons we chose to focus on patterns rather than individual numbers. With World3 we are engaged, we hope, in self-defeating prophecy.

To achieve our goals, we put into World3 the kinds of information you might use to understand the behavioral tendencies of thrown balls (or growing economies and populations), not the kinds of information you would need to describe the exact trajectory of one particular throw of one specific ball.

Our concern is with changes that unfold over many decades. So we focused our concern about pollution principally on the persistent materials—those that remain in the environment for many, many years. We represent persistent pollution as the collection of long-lived chemical compounds and metals generated by agriculture and industry that can affect the health of human beings and crops. We included a delay before pollution finds its way to a place where it can do measurable harm, because we know it takes time for a pesticide to work its way down into groundwater, or for a chlorofluorocarbon molecule to rise up and damage the ozone layer, or for mercury to wash into a river and accumulate in the flesh of fish. We represented the fact that natural processes can render most pollutants harmless after a while, but also the fact that those natural cleanup processes can themselves be impaired. Widely shared dynamic characteristics of persistent pollutants are included in World3, but the model does not distinguish among the unique features of PCBs, CFCs, DDT, heavy metals, and radioactive wastes.

In World3 we used the best numbers we could find, but we acknowledge a large range of uncertainty around many of our estimates. When there is doubt about important numbers, modelers test a wide range of possibilities. They look to see whether any estimates within the range of uncertainty lead to significantly different conclusions. For example, we made the best judgment we could from geologists' data about the amount of nonrenewable resources still remaining under the ground. Then we halved and doubled that number to see what difference it would make to the behavior of our model system if the geologists were wrong, or if we had misinterpreted their data.

Because of the uncertainties and simplifications we know exist in the model (and others that we suppose it must contain, though we have not yet recognized them), we do not put faith in the precise numerical path the model generates for population, pollution, capital, or food production. Still, we think the primary interconnections in World3 are good representations of the important causal mechanisms in human society. Those interconnections, not the precise numbers, determine the model's general behavior. As a consequence, we do have faith in the dynamic behaviors generated by World3. We will present 11 different scenarios for the future, through the year 2100, and we believe those scenarios substantiate important insights and principles about whether and under what conditions population, industry, pollution, and related factors in the future may grow, hold steady, oscillate, or collapse.

The Structure of World3

What are the primary interconnections? They begin with the feedback loops involving population and capital that we described in chapter 2. Those loops are reproduced in figure 4-4. They give population and capital the potential to grow exponentially, if the positive birth and investment loops dominate; the potential to decline, if the negative death and depreciation loops dominate; and the potential to stay constant, if the loops are balanced.

In all our feedback loop diagrams, such as figure 4-4, the arrows indicate simply that one variable influences another through physical or informational flows. You can tell the story of our assumptions by talking your way

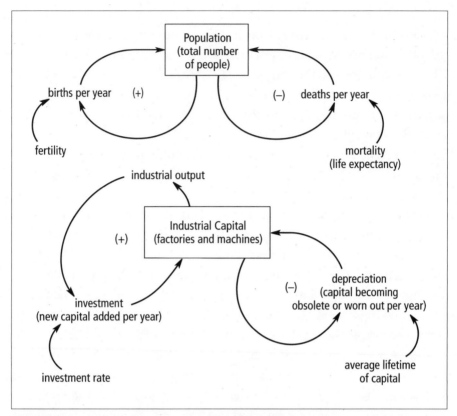

FIGURE 4-4 Feedback Loops Governing Population and Capital Growth
The central feedback loops of the World3 model govern the growth of population and of industrial capital. The two positive feedback loops involving births and investment generate the exponential growth
behavior of population and capital. The two negative feedback loops involving deaths and depreciation
tend to regulate this exponential growth. The relative strengths of the various loops depend on many
other factors in the system.

around each of our loops in turn. For example, "As industrial capital
increases, it will affect industrial output. Change in industrial output causes
investment to change. As investment changes, it will affect the stock of
industrial capital." The *nature* and *degree* of influence are not shown on the
diagrams, though of course they must be specified precisely in the mathematical equations that constitute World3. The direction of the influence
flows, clockwise or counterclockwise, make no difference at all. The
meaning is in composition of the loops.

The boxes in the diagrams indicate *stocks*. These may be important accumulations of physical quantities, such as population, factories, or pollution.

Or they may represent intangible accumulations, such as knowledge, aspirations, or technical capacity. The stocks in a system tend to change slowly, because they correspond to things or information with relatively long lifetimes. At each moment, the size of a stock represents the net effect over all past history of the rates flowing into the stock minus the rates flowing out. The factories in place, the number of people, the amount of pollutants, the quantity of nonrenewable resources remaining under the ground, the area of developed land—all these, and others, are important stocks in World3. They determine the limitations and possibilities of the model system at each moment of simulated time.

Feedback loops in the diagrams are marked with (+) if they are positive loops—self-reinforcing loops that can generate exponential growth or exponential decline. They are marked with (–) if they are negative loops—goal-seeking loops that reverse the direction of change or try to pull the system into balance or equilibrium.

Some of the ways population and capital influence each other in World3 are shown in figure 4-5. Industrial capital produces industrial output, which comprises many kinds of products, including those that are agricultural inputs, such as fertilizers, pesticides, and irrigation pumps. Agricultural inputs will be increased if food per person falls below desired food per person. The latter is a measure of market demand plus nonmarket programs to feed the population, and it changes with the society's level of industrialization. Agricultural inputs and the area of cultivated land help determine food production. Food is also affected by pollution, which comes from both industrial and agricultural activity. Both food per capita and pollution influence the mortality of the population.

Figure 4-6 shows the primary links connecting population, industrial capital, service capital, and nonrenewable resources in World3. Some industrial output takes the form of service capital—houses, schools, hospitals, banks, and the equipment they contain. This is invested in the service sector to raise the level of service capital. Output from service capital divided by the population gives the average level of services per person. Health services decrease the mortality of the population. Education and family planning services lower fertility and thus reduce the birth rate. Rising industrial output per capita also reduces fertility, an effect that results (after a delay)

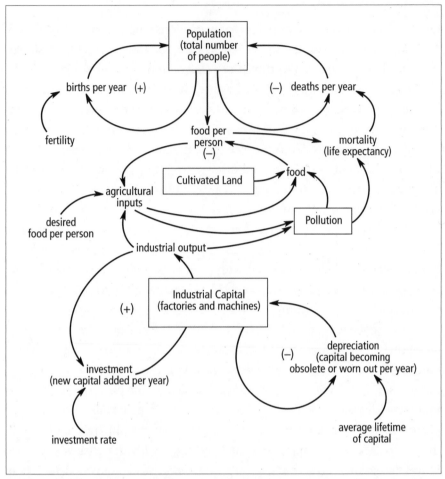

FIGURE 4-5 Feedback Loops of Population, Capital, Agriculture, and Pollution
Some of the interconnections between population and industrial capital operate through agricultural capital, cultivated land, and pollution. Each arrow indicates a causal relationship, which may be immediate or delayed, large or small, positive or negative, depending on the assumptions included in each model run.

from changing employment patterns. With industrialization, a society experiences increased costs of raising children and reduced benefits from large families. So the desired family size declines, and that lowers fertility.

Each unit of industrial output consumes nonrenewable resources. Technological advance in the model will gradually reduce the amount of resources needed per unit of industrial production, all else being equal. But the model does not allow industry to manufacture material goods out of nothing. As nonrenewable resources diminish, the efficiency of resource

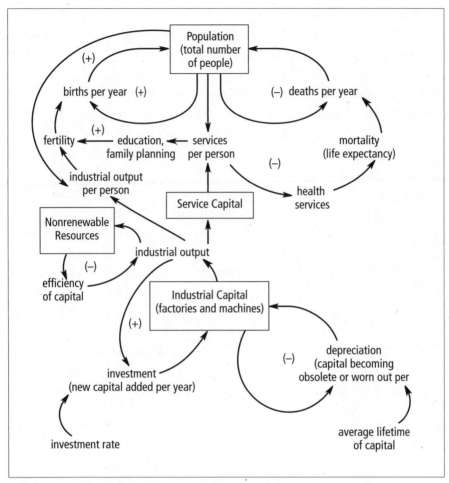

FIGURE 4-6 Feedback Loops of Population, Capital, Services, and Resources
Population and industrial capital are also influenced by the levels of service capital (such as health and education services) and of nonrenewable resources.

capital declines—one unit of capital delivers fewer and fewer resources to the industrial sector. As resources are consumed, the quality of the remaining reserves is assumed to decline. Deposits are assumed to be discovered deeper and to be exploited farther and farther from their places of use. That means more capital and energy will be necessary to extract, refine, and transport a ton of copper or a barrel of oil from the earth. Over the short term these trends may be offset by technological advance. Over the long term they will reduce the capacity for physical growth.

The relationship between resources remaining and the amount of capital required to obtain them is highly nonlinear. The general shape of the curve is shown in figure 4-7. This graph shows the energy necessary to extract and refine iron and aluminum at various ore grades. Energy is not capital (the actual amount of capital used in mining is hard to measure). But the amount of energy required to carry out a task gives important hints about the capital that is required. As the ore grade goes down, more rock must be lifted per ton of final resource; the rock must be crushed into finer particles; it must be sorted more accurately into its component minerals; and larger tailings piles must be dealt with. All of this requires machines. If more energy and capital are needed in the resource-producing sector, less investment is available for other purposes in the economy—all else being equal.

A diagram of all the interconnections in World3, illustrating all the assumptions incorporated in the model is reproduced on the World3 CD-ROM disk with much more detailed information about each of the 11 scenarios.

Still, it is not necessary to understand every one of these linkages in order to comprehend how the model works and to appreciate its scenarios. That requires only understanding of the model's most important features:

- The growth processes.
- The limits.
- The delays.
- The erosion processes.

We have already described the growth processes of population and capital in chapter 2. In chapter 3 we presented much information about environmental limits in the "real world." Next we will depict limits as they are represented in World3. Then we'll describe the delays and the erosion processes we have incorporated in our computer model.

The important question you should keep in mind throughout the following discussion is whether and under what conditions there are parallels or discrepancies between the computer model we are discussing and the "real" population and economy, as you know them through your own mental model. Where there are discrepancies, you will confront the ques-

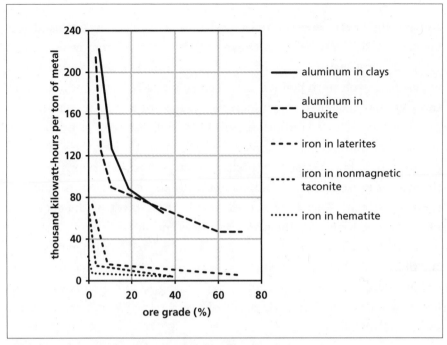

FIGURE 4-7 Energy Required to Produce Pure Metal from Ore
As their metal content declines, ores require increasingly large amounts of energy for their purification. (Source: N. J. Page and S. C. Creasey.)

tions that modelers face all the time. Which of the two models—yours or World3—seems to be more useful for thinking about the future? Is there a test that could help you find out? If the computer model seems more useful, which features of it must you incorporate into your mental model, so that your interpretations of global issues are useful and your actions are effective?

Limits and No Limits

An exponentially growing economy depletes resources, emits wastes, and diverts land from the production of renewable resources. As it operates within a finite environment, the expanding economy will begin to create stresses. These stresses begin to grow long before society arrives at the point where further growth is totally impossible. In response to the stresses, the environment begins to send signals to the economy. These signals take many forms. More energy is needed to pump water from diminishing aquifers, the investment required to develop a hectare of new farmland goes up, damage suddenly becomes apparent from emissions that were thought to be harmless, natural systems of the earth heal themselves more slowly under the assault from pollution. These rising real costs do not necessarily show up immediately through increased monetary prices, because market prices can be reduced by fiat or subsidies and distorted in other ways. Whether or not they are reinforced by rising market prices, the signals and pressures function as important parts of negative feedback loops. They seek to bring the economy into alignment with the constraints of the surrounding system. That is, they seek to stop the growth of the ecological footprint that is stressing the planet's sources and sinks.

World3 contains just a few kinds of limits related to planetary sources and sinks. (The "real world" contains many more.) All of them can be raised or lowered by technologies, actions, goal changes, and choices within the model world. In the standard, or default, version of World3, these are the source and sink limits:

- *Cultivated land* is the land used in all forms of agriculture. We assume the maximum possible value is 3.2 billion hectares. Cultivated land is expanded by investment in land development. As shown in figure 4-2, the cost of developing new land is assumed to rise as the most accessible and favorable land is developed first. Land is removed from cultivation both by erosion and by conversion to support urbanization and industrialization. Erosion may be reduced by investments in land maintenance.
- *Land fertility* is the inherent ability of soil to support plant growth—a combination of nutrient content, soil depth, water-holding capacity, climate, and soil structure. We assume that initial land fertility in 1900 was sufficient to produce each year 600 kilograms per hectare of grain-equivalent without the addition of fertilizers. Land fertility is degraded by pollution, which comes, in part, from industrial agricultural inputs. Degraded land left fallow is assumed to regain half its fertility in 20 years, or considerably faster if investment (such as manuring, planting legumes, or composting) is allocated to that purpose.

- *Yield achievable on each unit of land* depends on land fertility, air pollution, the intensity of industrial inputs such as fertilizer, and the level of technology. Industrial inputs increase the yield, but they have diminishing returns—each additional kilogram of fertilizer produces less additional yield than the kilogram before. We assume initially that the application of industrial inputs can enhance natural land fertility by a factor of 7.4 at most (notice this is 740 percent, and it applies to all land, not just the most productive fields!)—and we can test the uncertainty in this number by raising it even higher.

- *Nonrenewable resources* include minerals, metals, and fossil fuels. We typically start the model in the simulated year 1900 with a supply of these resources equal to more than 7,000 times the extraction rate in the year 1900.[7] The investment required to find and extract nonrenewable resources is assumed to rise as the richest and most convenient deposits are exploited first.

- The *ability of the earth to absorb pollution* is another limit represented in World3. This represents the net effect of many different processes that sequester or convert long-lived, toxic materials so that they can no longer cause damage. We are concerned here with materials such as organochlorines, greenhouse gases, and radioactive wastes. We express this limit as the assimilation half-life of the environment—the time required for natural processes to render harmless half the existing pollution. Of course, some toxic materials, such as isotopes of plutonium, have an almost infinite half-life. But we used very optimistic numbers here. We assumed that in 1970 the half-life was one year. If persistent pollution rises to 250 times the 1970 level, half-life would grow to 10 years. Quantitatively, this is the least understood limit of all, even for each individual pollutant. So there is enormous uncertainty about the size of this limit for persistent pollutants in combination.

Fortunately, the assumptions we make about the disappearance of persistent pollutants are not too important in the model, because those materials do not greatly affect the parameters in other parts of World3. We assumed that if accumulated pollution rises to five times what it was in 2000, it would reduce human life expectancy by less than 2 percent. In our 11 scenarios, persistent pollution seldom rises to five times its year-2000 value. When it does, in extreme scenarios, it reduces land fertility by 10 percent or more each year. But that decline can be offset by investments in land maintenance. We test other estimates in the model to see what their effects would be.

In the "real world" there are many other kinds of limits, including managerial and social ones. Some of them are implicit in the numbers in World3, since our model co-efficients came from the world's "actual" history over the past 100 years. But World3 has no war, no labor strikes, no corruption, no drug addiction, no crime, no terrorism. Its simulated population does its best to solve perceived problems, undistracted by struggles over political power or ethnic intolerance or by corruption. Since it lacks many social limits, World3 does paint an overly optimistic picture of future options.

What if we're wrong about, for example, the amount of nonrenewable resources under the ground remaining to be discovered? What if the actual number is only half of what we've assumed, or double, or 10 times more? What if the earth's "real" ability to absorb pollution without harm to the human population is not 10 times the 1990 rate of emissions, but 50 times or 500 times? (Or 0.5?) What if technologies are invented that decrease (or increase) pollution emission per unit of industrial production?

A computer model is a device for answering such questions. It can be used quickly and cheaply to conduct tests. All those "what ifs" are testable. It is possible, for example, to set the numbers on World3's limits astronomically high or to program them to grow exponentially. We have tried that. When all physical limits are effectively removed from the model system by an assumed technology that is unlimited in potential, practically instantaneous in impact, without cost, and error-free, the simulated human economy grows enormously. Figure 4-8, Scenario 0, shows what happens.

FIGURE 4-8 Scenario 0: Infinity In, Infinity Out

If all physical limits to the World3 system are removed, population peaks near 9 billion and starts a slow decline in a demographic transition. The economy grows until by the year 2080 it is producing 30 times the year-2000 level of industrial output, while using the same annual amount of nonrenewable resources and producing only one-eighth as much pollution per year.

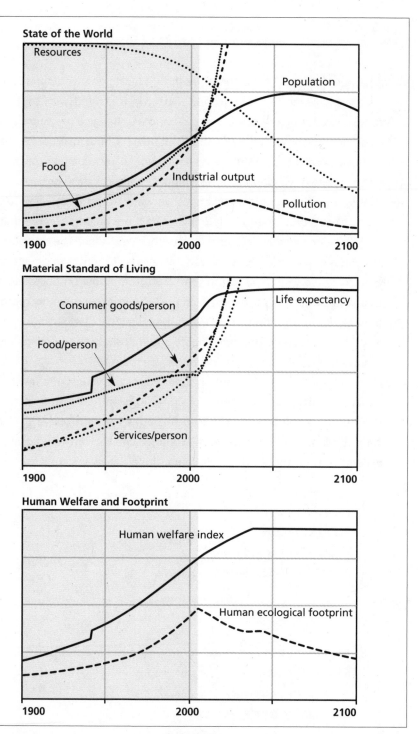

Scenario 0

How to Read World3 Scenarios

In chapters 4, 6, and 7 of this book we show 11 different "computer runs," or scenarios generated with World3. Each run is created with the same World3 model structure. But in each scenario we change a few numbers to test different estimates of "real world" parameters, or to incorporate more optimistic predictions about the development of technologies, or to see what happens if the world chooses different policies, ethics, or goals.

When we have made the changes we want to test in the new run, we instruct World3 to recalculate the interactions among its more than 200 equations as they influence each other over time. The computer calculates a new value for each variable every six months in simulated time from the year 1900 to the year 2100. The model produces more than 80,000 numbers for every scenario. There is no reason to reproduce all this information here. Few of the individual numbers have any meaning in isolation. Thus we simplify enormously, both to understand the model's results ourselves and to convey them to you.

We simplify by plotting out on time graphs the values of a few key variables, such as population, pollution, and natural resources. For this book we will provide three such graphs for each scenario. The format will be the same for each scenario. The top left-hand graph, called "State of the World," will show global totals for:

1. **Population.**
2. **Food production.**
3. **Industrial output.**
4. **Relative level of pollution.**
5. **Remaining nonrenewable resources.**

The middle graph, called "Material Standard of Living," will show average global values for:

6. **Food production per person.**
7. **Services per person.**
8. **Average life expectancy.**
9. **Consumption goods per person.**

The bottom graph, called "Human Welfare and Footprint," will show values for two global indicators:

10. **Human ecological footprint.**
11. **Human welfare index.**

All vertical scales start from zero. To facilitate the comparison, we have kept the vertical scale for each variable identical for all runs. We have, however, omitted the numerical values for the vertical scales for the variables, since their precise values at each point in simulated time are not meaningful. Note, furthermore, that variables on the same graph are each plotted on different scales with different units. For example, the scale for food per capita goes from 0 to 1,000 kilograms of grain-equivalents per person per year, while the scale for life expectancy goes from 0 to 90 years.

Since the numerical values are not significant, you should focus on how the shapes of the curves change from one scenario to another. Observe, however, in scenarios that portray collapse, we do not assign any meaning to the behavior of the curves beyond the point where they peak out and start to decline. Each scenario is portrayed fully out to the year 2100. But we do not describe the behavior of any model element after the point where one significant factor has started to collapse. Clearly a collapse of population or industry in the "real world" would change many important relationships and thereby invalidate many of the assumptions we have built into the model.

Each time we generate a scenario, the computer creates a detailed data table that gives the numerical value of every model variable every six months between the year 1900 and the year 2100. These tables give us enormous amounts of very detailed data. We see from the table on Scenario 0, for instance, that global population reaches a maximum value of 8,876,186,000 in model year 2065.0. The index of persistent pollution in this scenario rises from a value of 3.150530 in the year 2000 to reach its maximum value of 6.830552 in the model year 2026.5—it grows by a factor of 2.1680 over that period. But there is no useful information in most of those digits. No future number or date produced by our World3 warrants expression with five-digit precision. Remember, we are interested in the broad sweeps. We pay attention to a few key variables and ask only a few key questions. Which of the variables quit growing during the coming century? How rapidly do they grow or decline? What are the main factors that produce this behavior? Do the assumptions incorporated in a scenario cause a variable to grow faster or slower, to peak higher or lower? Which policy changes might produce a more favorable result?

When we convey to you the answers to those questions, scenario by scenario, we will simplify the computer reports greatly by using two rules. The timing of any maximum or minimum will be indicated only to the nearest decade end (we will round up from 5.0 to the next highest 10)—for example, not 2016, 2032.5, or 2035 but 2020, 2030, or 2040. Each value of a specific parameter and each ratio between two numerical values will be indicated only to the nearest significant digit. Therefore we would report to you the above information about Scenario 0 using the statement, "Global population reaches a maximum value of nine billion in model year 2070. The index of persistent pollution in this scenario rises from a value of three in the year 2000 to reach its maximum value of seven in the model year 2030— it grows by a factor of two over that period." These rules will sometimes appear to produce minor inconsistencies. Don't be distracted by them; they are due to rounding errors, and they do not have any influence on the central lessons we derive from the model.

The computer run in figure 4-8, Scenario 0, was produced by World3 after we changed its numbers to making the following assumptions:

- The amount of nonrenewable resources required to produce a unit of industrial output drops exponentially without limit at 5 percent per year, falling by 50 percent every 15 years as long as society strives to improve its resource efficiency.

- The amount of pollution generated per unit of industrial production drops exponentially without limit at 5 percent per year, when desired.
- The agricultural yield per unit of industrial input rises exponentially without limit at 5 percent per year, doubling every 15 years as long as society strives to increase its food production.
- All these technical achievements are effective throughout the world economy at no additional cost of capital and with an implementation delay of only 2 years (instead of 20 in the original model), once society has decided that such technology is desired.
- Human settlements encroach on agricultural land at one-fourth the rate normally assumed in World3, and humans experience no negative effects on their life expectancy from overcrowding.
- Agricultural output is no longer significantly reduced by pollution.

In this run, population slows its growth, levels off at almost nine billion, and then gradually declines, because the entire world population gets rich enough to experience the demographic transition. Average life expectancy stabilizes near 80 years worldwide. Average agricultural yield rises by the year 2080 to nearly six times its year-2000 value. Industrial output soars off the top of the graph—it is finally stopped at a very high level by a severe labor shortage, because there is 40 times as much industrial capital to manage and run as there was in the year 2000, but only 1.5 times as many people. (We could take away even that limit by assuming a sufficiently fast exponential rise in labor's capacity to use capital.)

By the simulated year 2080, the global economy is producing 30 times as much industrial output and 6 times as much food as it did in 2000. To achieve these results it has accumulated during the first eight decades of the twenty-first century almost 40 times as much industrial capital as it did during the entire twentieth century. While achieving that expansion in capital, the world portrayed in figure 4-8 reduces its nonrenewable resource use slightly and lowers its pollution emissions by a factor of eight compared with the year 2000. Human welfare increases 25 percent from 2000 to 2080, and the ecological footprint declines 40 percent. By the end of the scenario, the year 2100, the footprint is safely back below the sustainable level.

Some people believe in this kind of scenario; expect it; revel in it. We know stories of remarkable efficiency increases in particular countries or economic sectors or industrial processes. We mentioned many of these stories in chapter 3. We hope and believe that further efficiency improvements are possible, even 100-fold improvements. But the data presented in chapter 3 show no indication of the *whole global economy* achieving such gains so quickly. If nothing else would prevent such rapid changes, the lifetime of capital plants—the time it takes to replace or retrofit the vehicle fleet, building stock, and installed machinery of the global economy—and the ability of existing capital to produce that much new capital so fast make this "dematerialization" scenario unbelievable to us. The difficulties of achieving this infinity scenario would be magnified in "real life" by the many political and bureaucratic constraints preventing the price system from signaling that the needed technologies can be profitable.

We include this run here not because we think shows you a credible future of the "real world," but because we think it tells you something about World3 and something about modeling.

It reveals that World3 has built into its structure a self-limiting constraint on population and no self-limiting constraint on capital. The model is constructed in such a way that the global population will eventually level off and start declining, if industrial output per capita rises high enough. But we see little "real world" evidence that the richest people or nations ever lose interest in getting richer. Therefore, policies built into World3 represent the assumption that capital owners will continue to seek gains in their wealth indefinitely and that consumers will always want to increase their consumption. Those assumptions can and will be changed in policy runs presented in chapter 7.

Figure 4-8 also demonstrates one of the most famous principles of modeling: Garbage In, Garbage Out, or GIGO. If you put unrealistic assumptions into your model, you will get unrealistic results. The computer will tell you the logical consequences of your assumptions, but it will not tell you whether your assumptions are true. If you assume the economy can increase industrial capital accumulation 40-fold, that physical limits no longer apply, that technical changes can be built into the whole global capital plant in only two years without cost, World3 will give you virtually unlimited economic growth along with a declining ecological footprint. The important question about this and every other computer run is whether you believe the initial assumptions.

We don't believe the assumptions behind figure 4-8. We consider this a scenario that portrays an impossible technological utopia. So we label that run Infinity In, Infinity Out, or IFI-IFO (pronounced *iffy-iffo*). Under what we think are more "realistic" assumptions, the model begins to show the behavior of a growing system running into resistance from physical limits.

Limits and Delays

A growing physical entity will slow and then stop in a smooth accommodation with its limits (S-shaped growth) only if it receives accurate, prompt signals telling it where it is with respect to its limits, and only if it responds to those signals quickly and accurately (figure 4-9b).

Imagine that you are driving a car and up ahead you see a stoplight turn red. Normally you can halt the car smoothly just before the light, because you have a fast, accurate visual signal telling you where the light is, because your brain responds rapidly to that signal, because your foot moves quickly as you decide to step on the brake, and because the car responds immediately to the brake in a fashion you understand from frequent practice.

If your side of the windshield were fogged up and you had to depend on a passenger to tell you where the stoplight was, the short delay in communication could cause you to shoot past the light (unless you slowed down to accommodate the delay). If the passenger lied, or if you denied what you heard, or if it took the brakes two minutes to have an effect, or if the road had become icy, so that it unexpectedly took the car several hundred meters to stop, you would overshoot the light.

A system cannot come into an accurate and orderly balance with its limit if its feedback signal is delayed or distorted, if that signal is ignored or denied, if there is error in adapting, or if the system can respond only after a delay. If any of those conditions pertain, the growing entity will correct itself too late and overshoot (figures 4-9c & d).

We have already described some of the information and response delays in World3. One of them is the delay between the time when a pollutant is released into the biosphere and the time at which it does observable harm to human health or the human food supply. An example is the 10- to 15-year

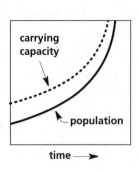

a) *Continuous growth results if*

- **Physical limits are very far off, or**

- **Physical limits are themselves growing exponentially.**

b) *Sigmoid growth results if*

- **Signals from physical limits to economy are instant, accurate, and responded to immediately, or**

- **The population or economy limits itself without needing signals from external limits.**

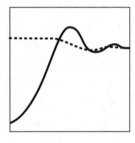

c) *Overshoot and oscillation results if*

- **Signals or responses are delayed, and**

- **Limits are unerodable or are able to recover quickly from erosion.**

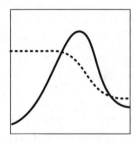

d) *Overshoot and collapse results if*

- **Signals or responses are delayed, and**

- **Limits are erodable (irreversibly degraded when exceeded).**

FIGURE 4–9 Structural Causes of the Four Possible Behavior Modes of the World3 Model

lag before a chlorofluorocarbon molecule released on the earth's surface begins to degrade the stratospheric ozone layer. Policy delays are also important. There is often a delay of many years between the date when a problem is first observed and the date when all important players agree on it and accept a common plan for action. These delays are described in the next chapter.

One illustration of these delays is provided by the percolation of PCBs through the environment. Since 1929 industry has produced some two million tons of the stable, oily, nonflammable chemicals called polychlorinated biphenyls, or PCBs.[8] They were used primarily to dissipate heat in electrical capacitors and transformers, but also as hydraulic fluid, lubricants, fire retardants, and constituents of paints, varnishes, inks, carbonless copy paper, and pesticides. For 40 years users of these chemicals dumped them in landfills, along roads, into sewers and water bodies, without thinking of the environmental consequences. Then in a landmark study in 1966, designed to detect DDT in the environment, Danish researcher Sören Jensen reported that in addition to DDT, he had found PCBs to be widespread as well.[9] Since then other researchers have found PCBs in almost all the globe's ecosystems.

> **PCBs are in almost every component of the global ecosystem. The hydrosphere is a major source of atmospheric PCBs. . . . PCB residues have also been detected in river, lake, and ocean sediments. . . . A comprehensive study of the Great Lakes ecosystem clearly illustrates the preferential bioconcentrations of PCB residues in the food chain.**
>
> Environment Canada, 1991

> **DDT and PCBs are the only organochlorines that have been monitored on a systematic basis in Arctic marine mammals. . . . The PCB levels in the breast milk of the Inuit women are among the highest ever reported. . . . A high consumption of fishes and sea mammals is probably the main route of intake for PCBs. . . . These results suggest that toxic compounds such as PCBs could play a role in the impairment of immunity and in the high occurrence of infection among Inuit children.**
>
> E. Dewailly, 1989

> **[In the Waddenzee on the coast of the Netherlands] the reproductive suc-
> cess of the seals receiving the diet with the highest level of [PCBs] was
> significantly decreased . . . [which shows that] the reproductive failure in
> common seals is related to feeding on fish from that polluted area. . . .
> These findings corroborate the results from experiments with mink,
> where PCBs impaired reproduction.**
>
> P.J.H. Reijnders, 1986

Most PCBs are relatively insoluble in water but soluble in fats, and they have very long lifetimes in the environment. They move quickly through the atmosphere, and slowly through soils or sediments in streams and lakes, until they are taken up into some form of life, where they accumulate in fatty tissue and increase in concentration as they move up the food chain. They are found in the greatest concentrations in carnivorous fish, seabirds and mammals, human fat, and human breast milk.

The impacts of PCBs on the health of humans and other animals are only slowly being revealed. The story is particularly difficult to unravel because PCB is a mixture of 209 closely related compounds, each of which may produce different effects. Nevertheless, it is becoming apparent that some PCBs act as endocrine disrupters. They mimic the action of some hormones, such as estrogen, and block the action of others, such as thyroid hormones. The effect—in birds, whales, polar bears, humans, any animal with an endocrine system—is to confuse delicate signals that govern metabolism and behavior. Especially in developing embryos, even minute concentrations of endocrine disrupters can wreak havoc. They can kill the developing organism outright, or they can impair its nervous system, intelligence, or sexual function.[10]

Because they migrate slowly, last a very long time, and accumulate in higher levels of a food chain, PCBs have been called a "biological time-bomb." Although PCB manufacture and use has been banned in many countries since the 1970s,[11] a huge stock still exists. Of the total amount of PCBs ever produced, much is still in use or stored in abandoned electrical equipment. In countries with hazardous waste laws, some of these old PCBs are being buried or disposed of by controlled incineration that breaks up their molecular structure and thus their bioactivity. In 1989 it was esti-

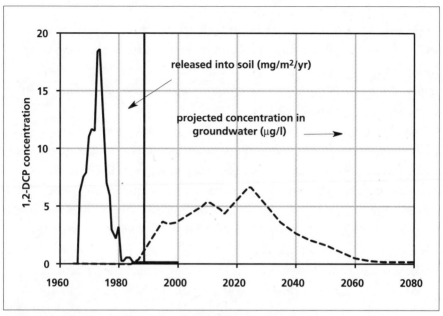

FIGURE 4-10 The Slow Percolation of 1,2-DCP into Groundwater
The soil disinfectant DCP was used heavily in the Netherlands in the 1970s, then it was restricted, and finally in 1990 it was banned. As a result, the concentration of DCP in the upper levels of agricultural soils has declined quickly. It was calculated in 1991 that its concentration in groundwater will not peak until 2020, however, and there will still be significant quantities of the chemical in the water after the middle of the twenty-first century. (Source: N. L. van der Noot.)

mated that 30 percent of all PCBs ever manufactured had already been released into the environment. Only 1 percent had reached the oceans. The 29 percent still unaccounted for was dispersed in soils, rivers, and lakes, where it would go on moving into living creatures for decades.[12]

Figure 4-10 shows another example of a pollution delay, the slow transport of chemicals through soil into groundwater. From the 1960s until 1990, when it was finally banned, the soil disinfectant 1,2-dichloropropene (DCPe) was applied heavily in the Netherlands in the cultivation of potatoes and flower bulbs. It contains a contaminant, 1,2-dichloropropane (DCPa), which, as far as scientists know, has an infinite lifetime in groundwater. A calculation for one watershed estimated that the DCPa already in the soil would work its way down into groundwater and appear there in significant concentrations only after the year 2010. Thereafter it was expected to contaminate the groundwater for at least a century in concentrations up to 50 times the European Union's drinking water standard.

The problem is not unique to the Netherlands. In the United States agricultural use of DCP was canceled in 1977. Yet the Washington State Pesticide Monitoring Program found the chemical at concentrations assumed to injure human health when it monitored ground water at 243 sites in 11 study areas between 1988 and 1995.[13]

A delay in a different sector of World3 is due to the population age structure. A population with a recent history of high birth rates contains many more young people than old people. Therefore, even if fertility falls, the population keeps growing for decades as the young people reach childbearing age. Though the number of children per family goes down, the number of families increases. Because of this "population momentum," if the fertility of the entire world population reaches replacement level (about two children per family on average) by the year 2010, the population will continue growing until 2060 and will level off at about eight billion.

There are many other delays in the "real world" system. Nonrenewable resources may be drawn down for generations before their depletion has serious economic consequences. Industrial capital cannot be built overnight. Once it is placed in operation, it has a lifetime of decades. An oil refinery cannot be converted easily or quickly into a tractor factory or a hospital. It even requires time to make it into a more efficient, less polluting oil refinery.

World3 has many delays in its feedback mechanisms, including all those mentioned above. We assume there is a delay between the release of pollution and its noticeable effect on the system. We assume a delay of roughly a generation before couples fully trust and adjust their decisions about family size to changing infant mortality rates. It normally takes decades in World3 before investment can be reallocated and new capital plant can be constructed and brought into full operation in response to a shortage of food or services. It takes time for land fertility to be regenerated or pollution to be absorbed.

The simplest and most incontrovertible physical delays are already sufficient to eliminate smooth sigmoid as a likely behavior for the world economic system. Because of the delays in the signals from nature's limits, overshoot is inevitable if there are no self-enforced limits. But that overshoot might, in theory, lead either to oscillation or to collapse.

Overshoot and Oscillation

If the warning signals from the limits to the growing entity are delayed, or if the response is delayed, and if the environment is not eroded when overstressed, then the growing entity will overshoot its limit for a while, make a correction, and under- shoot, then overshoot again, in a series of oscillations that usually damp down to an equilibrium within the limit (figure 4-9c).

Overshoot and oscillation can occur only if the environment suffers insignificant damage during periods of overload or can repair itself quickly enough to recover fully during periods of underload.

Renewable resources, such as forests, soils, fish, and rechargeable ground- water, are erodable, but they also have a self-regenerating capability. They can recover from a period of overuse, as long as it is not great enough or sus- tained enough that damage to the nutrient source, breeding stock, or aquifer is devastating. Given time, soil, seed, and a suitable climate, a forest can grow back. A fish stock can regenerate if its habitat and food supply are not destroyed. Soils can be rebuilt, especially with active help from farmers. Accumulations of many kinds of pollution can be reduced if the environ- ment's natural absorption mechanisms have not been badly disturbed.

Therefore the overshoot and oscillation behavior mode is a significant possibility for the world system. It has been demonstrated in some localities for some resources. New England, for example, has several times witnessed periods when more sawmills were built than could be supplied by the sus- tainable harvest of the region's forests. Each time that happened, the com- mercial timber stands were eventually depleted, mills had to be shuttered, and then the industry waited decades until the forest grew back and the overbuilding of sawmills could begin again. The coastal Norwegian fishery has gone through at least one cycle of fish depletion, with the government buying up and retiring fishing boats until the fish stocks could regenerate.

The decline phase of an overshoot and oscillation is not a pleasant period to live through. It can mean hard times for industries dependent upon an abused resource, or bad health in populations exposed to high pollution levels. Oscillations are best avoided. But they are not usually fatal to a system.

Overshoots can become catastrophic when the damage they cause is irreversible. Nothing can bring back an extinct species. Fossil fuels are permanently destroyed in the very act of using them. Some pollutants, such as radioactive materials, can't be rendered harmless by any natural mechanism. If the climate is significantly altered, geological data suggest that temperature and precipitation patterns probably will not return to normal within a time period meaningful to human society. Even renewable resources and pollution absorption processes can be permanently destroyed by prolonged or systematic misuse. When tropical forests are cut down in ways that preclude their regrowth, when the sea infiltrates freshwater aquifers with salt, when soils wash away leaving only bedrock, when a soil's acidity is changed sufficiently to flush out the heavy metals it has stored, then the earth's carrying capacity is diminished permanently, or at least for a period that appears permanent to human beings.

Therefore, the overshoot and oscillation mode is not the only one that could be manifested as humanity approaches the limits to growth. There is one more possibility.

Overshoot and Collapse

If the signal or response from the limit is delayed and if the environment is irreversibly eroded when overstressed, then the growing economy will overshoot its carrying capacity, degrade its resource base, and collapse (figure 4-9d).

The result of overshoot and collapse is a permanently impoverished environment and a material standard of living much lower than what would have been possible if the environment had never been overstressed.

The difference between the overshoot and oscillation and overshoot and collapse is the presence of *erosion loops* in a system. These are positive feedback loops of the worst kind. Normally they are dormant, but when a situation gets bad, they make it worse by carrying a system downward at an ever-increasing pace.

For example, grasslands all over the world have co-evolved with grazing animals such as buffalo, antelope, llamas, or kangaroos. When grasses are eaten down, the remaining stems and roots extract more water and nutrients from the soil and send up more grass. The number of grazers is held in check by predation, seasonal migration, and disease. The ecosystem does not erode. But if the predators are removed, the migrations are stymied, or the land is overstocked, an overpopulation of grazers can eat the grass down to the roots. That can precipitate rapid erosion.

The less vegetation there is, the less cover there is for the soil. With loss of cover, the soil begins to blow away in the wind or wash away in the rain. The less soil there is, the less vegetation can grow. Loss of vegetation allows still more soil to erode away. And so on. Land fertility spirals down until the grazing range has become a desert.

There are several erosion loops in World3. For example:

- If people become more hungry, they work the land more intensively. This produces more food in the short term at the expense of investments in long-term soil maintenance. Lower soil fertility then brings food production down even farther.
- When problems appear that require more industrial output—pollution that requires abatement equipment, for example, or hunger that calls for more agricultural inputs, or resource shortages that stimulate the discovery and processing of new resources—available investment may be allocated to solving the immediate problem, rather than maintaining existing industrial capital against depreciation. If the established industrial capital plant begins to decline, that makes even less industrial output available in the future. Reductions in output can lead to further postponed maintenance and further decline in the industrial capital stock.
- In a weakening economy, services per capita may decline. Reduced expenditures on family planning can eventually cause birth rates to increase. This produces growth in the population, which lowers services per capita even farther.
- If pollution levels increase too much, they can erode the pollution absorption mechanisms themselves, reducing the rate of pollution assimilation and raising the rate of pollution buildup still more.

This last erosive mechanism, impairment of the natural mechanisms for pollution assimilation, is particularly insidious. It is a phenomenon for which we had little evidence when we first designed World3 more than 30 years ago. At the time we had in mind such interactions as dumping pesticides into water bodies, thereby killing the organisms that normally clean up organic wastes; or emitting both nitrogen oxides and volatile organic chemicals into the air, which react with each other to make more damaging photochemical smog.

Since then other examples of the degradation of the earth's pollution control devices have come to light. One of them is the apparent ability of short-term air pollutants, such as carbon monoxide, to deplete scavenger hydroxyl radicals in the air. These hydroxyl radicals normally react with and destroy the greenhouse gas methane. When air pollution removes them from the atmosphere, methane concentrations increase. By destroying a pollution cleanup mechanism, short-term air pollution can make long-term climate change worse.[14]

Another such process is the ability of air pollutants to weaken or kill forests, thereby diminishing a sink for the greenhouse gas carbon dioxide. A third is the effect of acidification—from fertilizers or industrial emissions—on soils. At normal levels of acidity, soils are pollution absorbants. They bind with and sequester toxic metals, keeping them out of streams and groundwater and thus out of living organisms. But these bonds are broken under acidic conditions. W. M. Stigliani described this process in 1991.

> As soils acidify, toxic heavy metals, accumulated and stored over long time periods (say, decades to a century) may be mobilized and leached rapidly into ground and surface waters or be taken up by plants. The ongoing acidification of Europe's soils from acid deposition is clearly a source of real concern with respect to heavy metal leaching.[15]

Besides the ones we included in World3, there are many other positive feedback loops in the "real world" with the potential to produce rapid erosion. We have mentioned the potential for erosion in physical and biological systems. An example of a very different kind would be breakdown in the social order. When a country's elites believe it is acceptable to have large differentials in well-being within their nation, they can use their power to produce big differences in income between themselves and most of the citizenry. This inequality can lead the middle classes to frustration, anger, and protests. The disruption that results from protests may lead to repression. Exercising force isolates the elites even farther from the masses and amplifies among the powerful the ethics and values that justify large gaps between them and the majority of the population. Income differentials rise, anger and frustration grow, and this can call forth even more repression. Eventually there may be revolution or breakdown.

It is difficult to quantify erosive mechanisms of any sort, because erosion is a whole-system phenomenon having to do with interactions among multiple forces. It appears only at times of stress. By the time it becomes obvious, it isn't easily stopped. But despite these uncertainties, we can say confidently that any system containing a latent erosion process also contains the possibility of collapse, if it is overstressed.

On a local scale, overshoot and collapse can be seen in the processes of desertification, mineral or groundwater depletion, poisoning of agricultural soils or forest lands by long-lived toxic wastes, and extinction of species. Abandoned farms, deserted mining towns, and forsaken industrial dumps all testify to the "reality" of this system behavior. On a global scale, overshoot and collapse could mean the breakdown of the great supporting cycles of nature that regulate climate, purify air and water, regenerate biomass, preserve biodiversity, and turn wastes into nutrients. When we first published our results in 1972, the majority of people thought human disruption of natural processes on a global scale was inconceivable. Now it is the subject of newspaper headlines, the focus of scientific meetings, and the object of international negotiations.[16]

World3: Two Possible Scenarios

In the simulated world of World3, the primary goal is growth. The World3 population will stop growing only when it is very rich. Its economy will stop growing only when it runs into limits. Its resources decline and deteriorate with overuse. The feedback loops that connect and inform its decisions contain substantial delays, and its physical processes have considerable momentum. It should therefore come as no surprise that the most likely mode of behavior of the model world is overshoot and collapse.

The graphs in figure 4-11, Scenario 1, show the behavior of World3 when it is run "as is," with numbers we consider a "realistic" description of the situation as it appeared on average during the latter part of the twentieth century, with no unusual technical or policy assumptions. In 1972 we called it the "standard run." We did not consider it to be the most probable future, and we certainly didn't present it as a prediction. It was just a place to start, a base for comparison. But many people imbued the "standard run" with more importance than the scenarios that followed. To prevent that from happening again, we'll just call it "a reference point" and refer to each scenario by number; this is Scenario 1.

In Scenario 1 the society proceeds along a very traditional path as long as possible without major policy change. It traces the broad outline of history as we know it throughout the twentieth century. The output of food, industrial goods, and social services increases in response to obvious needs and subject to the availability of capital. There is no extraordinary effort, beyond what makes immediate economic sense, to abate pollution, conserve resources, or protect the land. This simulated world tries to bring all people through the demographic transition and into a prosperous industrial economy. The world in Scenario 1 acquires widespread health care and birth control as the service sector grows; it applies more agricultural inputs and gets higher yields as the agricultural sector grows; it emits more pollutants, demands more nonrenewable resources, and becomes capable of greater production as the industrial sector grows.

The population in Scenario 1 rises from 1.6 billion in the simulated year 1900 to 6 billion in the year 2000 and more than 7 billion by 2030. Total industrial output expands by a factor of almost 30 between 1900 and 2000 and then by 10 percent more by 2020. Between 1900 and 2000 only about 30 percent of the earth's total stock of nonrenewable resources is used; more than 70 percent of these resources remain in 2000. Pollution levels in the

FIGURE 4-11 Scenario 1: A Reference Point

The world society proceeds in a traditional manner without any major deviation from the policies pursued during most of the twentieth century. Population and production increase until growth is halted by increasingly inaccessible nonrenewable resources. Ever more investment is required to maintain resource flows. Finally, lack of investment funds in the other sectors of the economy leads to declining output of both industrial goods and services. As they fall, food and health services are reduced, decreasing life expectancy and raising average death rates.

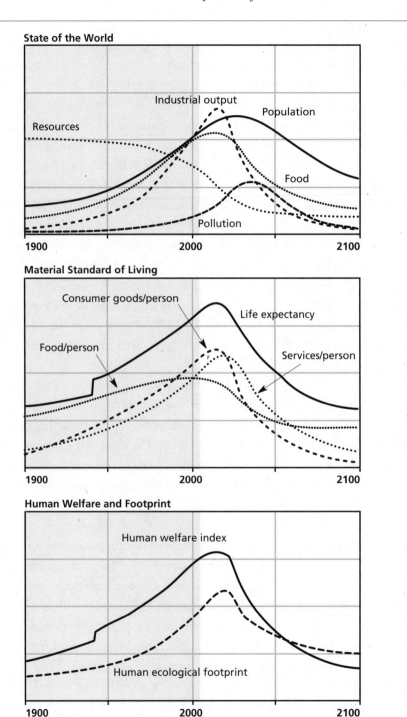

State of the World

Industrial output

Population

Resources

Food

Pollution

1900　　　　　2000　　　　　2100

Material Standard of Living

Consumer goods/person

Life expectancy

Food/person

Services/person

1900　　　　　2000　　　　　2100

Human Welfare and Footprint

Human welfare index

Human ecological footprint

1900　　　　　2000　　　　　2100

Scenario 1

simulated year 2000 have just begun to rise significantly, to 50 percent above the 1990 level. Consumer goods per capita in 2000 are 15 percent higher than in 1990, and nearly eight times higher than in 1900.[17]

If you cover the right half of the graphs in Scenario 1, so you can see only the curves up to the year 2000, the simulated world looks very successful. Life expectancy is increasing, services and goods per capita are growing, total food production and industrial production are rising. Average human welfare is increasing continuously. A few clouds do appear on the horizon: Pollution levels are rising, and so is the human ecological footprint. Food per person is stagnating. But generally the system is still growing, with few indications of the major changes just ahead.

Then suddenly, a few decades into the twenty-first century, the growth of the economy stops and reverses rather abruptly. This discontinuation of past growth trends is principally caused by rapidly increasing costs of nonrenewable resources. This cost rise works its way through the various economic sectors in the form of increasingly scarce investment funds. Let's follow the process.

In the simulated year 2000, the nonrenewable resources remaining in the ground would have lasted 60 years at the year-2000 consumption rate. No serious resource limits are then in evidence. But by 2020 the remaining resources constitute only a 30-year supply. Why does this shortage arise so quickly? It occurs because growth in industrial output and population raise resource consumption while drawing down the resource stock. Between 2000 and 2020 population increases by 20 percent and industrial output by 30 percent. During those two decades in Scenario 1, the growing population and industrial plant use nearly the same amount of nonrenewable resources as the global economy used in the entire century before! Naturally, more capital is then required to find, extract, and refine what nonrenewables remain—in the incessant effort of the simulated world to fuel further growth.

As nonrenewable resources become harder to obtain in Scenario 1, capital is diverted to producing more of them. That leaves less industrial output to invest in sustaining the high agricultural output and further industrial growth. And finally, around 2020, investment in industrial capital no longer keeps up with depreciation. (This is *physical* investment and depreciation; in other words, wear and tear and obsolescence, not monetary depreciation in accounting books.) The result is industrial decline, which is hard to avoid in this situation, since the economy cannot stop putting capital into the resource sector. If it did, the scarcity of materials and fuels would restrict industrial production even more quickly.

So maintenance and upkeep are deferred, the industrial plant begins to decline, and along with it go the production of the various industrial outputs that are necessary to maintain growing capital stocks and production rates in the other sectors of the economy. Eventually the declining industrial sector forces declines in the service and agricultural sectors, which depend on industrial inputs. The decline of industry has an especially serious impact on agriculture in Scenario 1, since land fertility has already been degraded somewhat by overuse prior to the year 2000. Consequently, food production is maintained mainly by compensating for this degradation with industrial inputs such as fertilizer, pesticides, and irrigation equipment. Over time the situation grows increasingly serious, because the population keeps rising due to lags inherent in the age structure and in the process of social adjustment to fertility norms. Finally, about the year 2030, population peaks and begins to decrease as the death rate is driven upward by lack of food and health services. Average life expectancy, which was 80 years in 2010, begins to decline.

This scenario portrays a "nonrenewable resource crisis." It is *not a prediction*. It is not meant to forecast precise values of any of the model variables, nor the exact timing of events. We do not believe it represents the most likely "real world" outcome. We'll show another possibility in a moment, and many more in chapters 6 and 7. The strongest statement we can make about Scenario 1 is that it portrays the likely *general behavior mode* of the system, *if* the policies that influence economic growth and population growth in the future are similar to those that dominated the last part of the twentieth century, *if* technologies and values continue to evolve in a manner representative of that era, and *if* the uncertain numbers in the model are roughly correct.

What if our assumptions and numbers aren't correct? What difference would it make if, for example, there are actually twice as many nonrenewable resources waiting to be discovered in the ground as we assumed in Scenario 1? That test is shown in figure 4-12, Scenario 2.

As you can see, resource depletion occurs considerably later in this run than it did in Scenario 1, allowing growth to continue longer. Expansion continues for an additional 20 years, long enough to achieve one more doubling in industrial output and resource use. The population also grows longer, reaching a peak of more than eight billion in the simulated year 2040. Despite these extensions, the general behavior of the model is still overshoot and collapse. The collapse now comes primarily from intense pollution of the global environment.

Higher levels of industrial output cause pollution to grow immensely; the pollution level in Scenario 2 peaks about 50 years later than it does in Scenario 1, at a level around five times higher. Part of this rise is due to greater pollution generation rates, and part is due to the fact that pollution assimilation processes are becoming impaired. At the peak around 2090, the average lifetime of pollutants in the environment has more than tripled from its year-2000 value. Huge inputs of fertilizer, pesticides, and other agricultural inputs further add to the ecological footprint.

The pollution has a major impact on land fertility, which declines dramatically in Scenario 2 through the first half of the twenty-first century. Even with increased investments to combat that loss, the effect on land fertility restoration is not sufficient to prevent yield and food production from falling sharply after 2030. So death rates rise. Still more capital is allocated to the agriculture sector in a vain attempt to stop hunger, and eventually the industrial sector stops growing due to lack of reinvestment.

Scenario 2 portrays a "global pollution crisis." Into the first half of the twenty-first century, pollution levels rise sufficiently to affect the fertility of the land. This could happen in the "real world" through soil contamination by heavy metals or persistent chemicals, through climate change altering growth patterns faster than farmers can adapt, or through increased ultraviolet radiation from a diminished ozone layer. Land fertility declines only

FIGURE 4-12 Scenario 2: More Abundant Nonrenewable Resources
If we double the nonrenewable resource endowment assumed in Scenario 1, and furthermore postulate that advances in resource extraction technologies are capable of postponing the onset of increasing extraction costs, industry can grow 20 years longer. Population peaks at 8 billion in 2040, at much higher consumption levels. But pollution levels soar (outside the graph!), depressing land yields and requiring huge investments in agricultural recovery. The population finally declines because of food shortages and negative health effects from pollution.

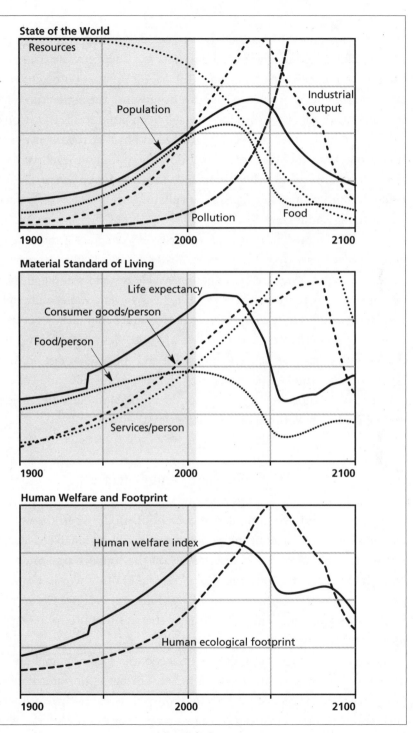

State of the World

Resources

Population

Industrial output

Pollution   Food

1900   2000   2100

Material Standard of Living

Life expectancy

Consumer goods/person

Food/person

Services/person

1900   2000   2100

Human Welfare and Footprint

Human welfare index

Human ecological footprint

1900   2000   2100

Scenario 2

slightly between 1970 and 2000, but it goes down 20 percent from 2000 to 2030, and by 2060 the inherent land fertility is a fraction of its 2000 value. At the same time, soil erosion is high. Total food production begins to fall in 2030, causing the economy to shift investment into the agriculture sector to maintain food adequacy. But the damage from pollution is too strong, and food production never recovers. In the second half of the twenty-first century, along with the effect of scarce food, the pollution level rises so high that it forces the average life expectancy to much lower levels. The human ecological footprint is huge, until the collapse shrinks it toward values similar to the previous century.

Which is a more likely future, Scenario 1 or Scenario 2? If there were a scientific way of answering that question, it would depend on evidence about the "actual" amount of undiscovered nonrenewable resources in the ground. But we cannot know those numbers for sure. In any event, there are many more uncertain numbers to test, and many technical and policy changes to try. We'll come to them in chapters 6 and 7. All that World3 has told us so far is that the model system has a tendency to overshoot and collapse. In fact, in the thousands of model runs we have tried over the years, overshoot and collapse has been by far the most frequent—but not the inevitable—outcome. By now the reasons for that behavior should be quite clear.

Why Overshoot and Collapse?

A population and economy are in overshoot mode when they are withdrawing resources or emitting pollutants at an unsustainable rate, but are not yet in a situation where the stresses on the support system are strong enough to reduce the withdrawal or emission. In other words: Humanity is in overshoot when the human ecological footprint is above the sustainable level, but not yet large enough to trigger changes that produce a decline in its ecological footprint.

Overshoot comes from delays in feedback. Decision makers in the system do not immediately get, or believe, or act upon information that limits have been exceeded. Overshoot is possible because there are accumulated resource stocks that can be drawn down. For example, you can spend more each month than you earn, at least for a while, if you have stored up funds in a bank account. You can drain water out of a bathtub faster than it is replenished by the faucet, at least until you have exhausted the initial stock of water in the tub. You can remove from a forest wood exceeding its annual growth rate as long as you start with a standing stock of wood that has grown and accumulated over many decades. You can build up enough herds to overgraze, or boats to overfish, if you have initially accumulated stocks of forage and fish that were not exploited in the past. The larger the initial stocks, the higher and longer the overshoot can be. If a society takes its signals from the simple availability of stocks, rather than from their rates of replenishment, it will overshoot.

Physical momentum adds to the delay in the warning signals, and it is another source of delay in the response to them. Because of the time it takes forests to regrow, populations to age, pollutants to work their way through the ecosystem, polluted waters to become clean again, capital plants to depreciate, people to be educated or retrained, the system can't change overnight, even after it perceives and acknowledges the problems. To steer correctly, a system with inherent momentum needs to be looking ahead at least as far as its momentum can carry it. The longer it takes a boat to turn, the farther ahead its radar must see. The political and market systems of the globe do not look far enough ahead.

The final contributor to overshoot is the pursuit of growth. If you were driving a car with fogged windows or faulty brakes, the first thing you would do to avoid overshoot would be to *slow down*. You would certainly not insist on accelerating. Delays in feedback can be handled as long as the system is not moving too fast to receive signals and respond before it hits the limit. Constant acceleration will take any system, no matter how clever and farsighted and well designed, to the point where it can't react in time. Even a car and driver functioning perfectly are unsafe at high speeds. The faster the growth, the higher the overshoot, and the farther the fall. The political and economic systems of the globe are dedicated to achieving the highest possible growth rates.

What finally converts overshoot to collapse is erosion, aided by nonlinearities. Erosion is a stress that multiplies itself if it is not quickly remedied. Nonlinearities like the ones shown in figures 4-2 and 4-7 are equivalent to *thresholds*, beyond which a system's behavior suddenly changes. A nation can mine copper ore down to lower and lower grades, but below a certain grade mining costs suddenly escalate. Soils can erode with no effect on crop yield until the soil becomes more shallow than the root zone of the crop. Then further erosion leads rapidly to desertification. The presence of thresholds makes the consequences of feedback delays even more serious. If you're driving that car with the fogged windows and faulty brakes, sharp curves mean you need to go even more slowly.

Any population–economy–environment system that has feedback delays and slow physical responses; that has thresholds and erosive mechanisms; and that grows rapidly is literally *unmanageable*. No matter how fabulous its technologies, no matter how efficient its economy, no matter how wise its leaders, it can't steer itself away from hazards. If it constantly tries to accelerate, it will overshoot.

By definition, overshoot is a condition in which the delayed signals from the environment are not yet strong enough to force an end to growth. How, then, can society tell if it is in overshoot? Falling resource stocks and rising pollution levels are the first clues. Here are some other symptoms:

- Capital, resources, and labor diverted to activities compensating for the loss of services that were formerly provided without cost by nature (for example, sewage treatment, air purification, water purification, flood control, pest control, restoration of soil nutrients, pollination, or the preservation of species).
- Capital, resources, and labor diverted from final goods production to exploitation of scarcer, more distant, deeper, or more dilute resources.
- Technologies invented to make use of lower-quality, smaller, more dispersed, less valuable resources, because the higher-value ones are gone.
- Failing natural pollution cleanup mechanisms; rising levels of pollution.
- Capital depreciation exceeding investment, and maintenance deferred, so there is deterioration in capital stocks, especially long-lived infrastructure.
- Growing demands for capital, resources, and labor used by the

military or industry to gain access to, secure, and defend resources that are increasingly concentrated in fewer, more remote, or increasingly hostile regions.

- Investment in human resources (education, health care, shelter) postponed in order to meet immediate consumption, investment, or security needs, or to pay debts.
- Debts a rising percentage of annual real output.
- Eroding goals for health and environment.
- Increasing conflicts, especially conflicts over sources or sinks.
- Shifting consumption patterns as the population can no longer pay the price of what it really wants and, instead, purchases what it can afford.
- Declining respect for the instruments of collective government as they are used increasingly by the elites to preserve or increase their share of a declining resource base.
- Growing chaos in natural systems, with "natural" disasters more frequent and more severe because of less resilience in the environmental system.

Do you observe any of these symptoms in your "real world"? If you do, you should suspect that your society is in advanced stages of overshoot.

A period of overshoot does not necessarily lead to collapse. It does require fast and determined action, however, if collapse is to be avoided. The resource base must be protected quickly, and the drains on it sharply reduced. Excessive pollution levels must be lowered, and emission rates reduced back to levels below what is sustainable. It may not be necessary to reduce population or capital or living standards. What must go down quickly are material and energy throughputs. In other words, the ecological footprint of humanity must be lowered. Fortunately (in a perverse way), there is so much waste and inefficiency in the current global economy that there is tremendous potential for reducing the footprint while still maintaining or even raising the quality of life.

In summary, here are the central assumptions in the World3 model that give it a tendency to overshoot and collapse. *If you wish to disagree with our model, our thesis, our book, or our conclusions, these are the points to contest:*

- Growth in the physical economy is considered desirable; it is central to our political, psychological, and cultural systems. Growth of both the population and the economy, when it does occur, tends to be exponential.
- There are physical limits to the sources of materials and energy that sustain the population and economy, and there are limits to the sinks that absorb the waste products of human activity.
- The growing population and economy receive signals about physical limits that are distorted, noisy, delayed, confused, or denied. Responses to those signals are delayed.
- The system's limits are not only finite, but erodable when they are overstressed or overused. Furthermore, there are strong non-linearities—thresholds beyond which damage rises quickly and can become irreversible.

Listing these causes of overshoot and collapse also gives a list of ways to avoid them. To change the system so that it is sustainable and manageable, the same structural features have to be reversed:

- Growth in population and capital must be slowed and eventually stopped by human decisions enacted in anticipation of future problems rather than by feedback from external limits that have already been exceeded.
- Throughputs of energy and materials must be reduced by drastically increasing the efficiency of capital. In other words, the ecological footprint must be reduced through dematerialization (less use of energy and materials to obtain the same output), increased equity (redistribution from the rich to the poor of the benefits from using energy and materials), and lifestyle changes (lowering demands or shifting consumption towards goods and services that have fewer negative impacts on the physical environment).
- Sources and sinks must be conserved and, where possible, restored.
- Signals must be improved and reactions speeded up; society must

look farther ahead and base current actions on long-term costs and benefits.

- Erosion must be prevented and, where it already exists, slowed and then reversed.

In chapters 6 and 7 we will show how these sorts of changes can alter the tendency toward overshoot and collapse of the World3 system—and, we believe and hope, the world. But first, in chapter 5, we take a short digression for a story that illustrates all the dynamic principles we have presented in this chapter—and it is one story that offers a foundation for hope.

CHAPTER 5

Back from Beyond the Limits: The Ozone Story

We find ourselves, one way or another, in the midst of a large-scale
experiment to change the chemical construction of the stratosphere,
even though we have no clear idea of what the biological or meteorological consequences may be.

—F. SHERWOOD ROWLAND, 1986

In this chapter we offer a useful story about shooting past an important
limit, observing the consequences, and then struggling with great success
to bring human activity back down to sustainable levels. The story concerns
the limited capacity of the stratospheric ozone layer to absorb human-made
chlorofluorocarbon chemicals (CFCs).[1] The final chapter of our story will
not be written for at least several more decades. But so far this narrative
gives the basis for hope. It shows that people and institutions, despite
common human failings, can come together on a global scale, diagnose an
overshoot problem, then design and implement solutions. In this case the
global society will sacrifice relatively little for accepting the necessity of
living within a limit.

The principal events of the ozone story are the following. Scientists
sounded the first warnings about the disappearing ozone layer and then
organized across political boundaries to develop an effective research effort.
But they could do so only after they managed to get beyond their own per-
ceptual blinders and their inexperience with the political process.
Consumers organized quickly to reverse a harmful trend, but their actions
alone were not enough for a permanent solution to the problem.
Governments and corporations at first acted as footdraggers and naysayers,
but then some of them emerged as courageous and selfless leaders.
Environmentalists were labeled as wild-eyed alarmists, but in this case they
turned out to have underestimated the problem.

The United Nations in this story showed its capacity for passing crucial information around the world and providing neutral ground and sophisticated facilitation, as governments worked through an undeniably international problem. Industrializing nations found in the ozone crisis a new power to act on their own behalf, by refusing to cooperate until they were guaranteed the technical and financial support they badly needed.

In the end the world's nations acknowledged that they had overrun a serious limit. Soberly, reluctantly, they agreed to give up a set of profitable and useful industrial products. They started to do so before any economic, biological, or human damage was observed, and before there was complete scientific certainty. They probably did it in time.

The Growth

First invented in 1928, chlorofluorocarbons (CFCs) are some of the most useful compounds ever synthesized by human beings. They do not seem to poison any living things, perhaps because they are so chemically stable. They do not burn or react with other substances or corrode materials. They have low thermal conductivity, so they make excellent insulators when blown into plastic foam for hot-drink cups, hamburger containers, or wall insulation. Some CFCs evaporate and recondense at room temperatures. This property makes them perfect coolants for refrigerators and air conditioners. (In that use they are often known under the trade name Freon). CFCs make good solvents for cleaning metals, from the intricate microspaces on electronic circuit boards to the rivets that hold airplanes together. CRCs are inexpensive to make, and they can be discarded safely—or so everyone thought—simply by releasing them as gases into the atmosphere or by burying the products that contain them in landfills.

As figure 5-1 shows, from 1950 to 1975 world production of CFCs grew at more than 11 percent per year—almost doubling every six years. By the mid-1980s industry was manufacturing a million tons of CFCs annually. In the United States alone, CFC coolants were at work in 100 million refrigerators, 30 million freezers, 45 million home air conditioners, 90 million car air conditioners, and hundreds of thousands of coolers in restaurants, super-

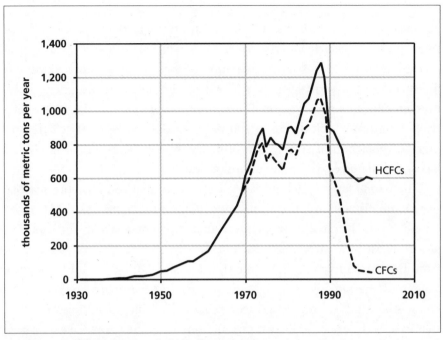

FIGURE 5-1 World Production of Chlorofluorocarbons
Production of CFCs grew rapidly until 1974, when the first papers postulating their effects on the ozone layer appeared. The subsequent decrease was due to environmental activism against CFC-containing aerosol spray cans, which were banned in the United States in 1978. After 1982 the expansion of other CFC uses caused production to rise again temporarily. In 1990 production started to drop as the internationally agreed-upon phaseout of CFCs began. HCFCs are still allowed as substitutes; the phaseout of this class of chemicals is slated for 2030 through 2040. (Source: Alternative Fluorocarbons Environmental Acceptability Study.)

markets, and refrigerated trucks.[2] The average North American or European was using 0.9 kilogram (two pounds) of CFCs per year. The average resident of China or India was using less than 0.03 kg (one ounce).[3] For a number of chemical companies in North America, Europe, Russia, and Asia these substances were a major source of income. Thousands of companies considered them essential inputs to their own production processes.

The Limit

The hero of our story is an invisible gas called ozone—three oxygen atoms stuck together (O_3), as opposed to ordinary oxygen, which consists of two bound oxygen atoms (O_2). Ozone is so reactive that it attacks and oxidizes almost anything it contacts. The lower atmosphere offers ozone a dense population of particles and surfaces with which it can react. Of special interest are plant tissues and human lungs. Near the earth's surface ozone is a destructive but short-lived air pollutant. High in the stratosphere, however, an ozone molecule normally encounters very little, and it lasts a relatively long time, typically 50 to 100 years. Ozone is constantly created on high by the action of sunlight on ordinary oxygen. Thus an "ozone layer" has accumulated 6 to 20 miles above the earth.

The ozone layer is rich in ozone only in comparison with the scarcity of that gas elsewhere in the atmosphere. Only 1 molecule in 100,000 in the ozone layer is actually ozone. But that concentration is sufficient to absorb most of a particularly harmful ultraviolet wavelength called UVB from the sun's incoming light (see figure 5-2). UVB light is a hail of little bullets of energy just the right frequency to take apart organic molecules—the kinds of molecules that make up all life, including the DNA that carries the code for life's reproduction. Thus the ozone layer is a gossamer veil with a crucial function.

When living organisms are hit by UVB, one possible result is cancer. UVB light has long been known to cause skin cancer in laboratory animals. Nearly all human skin cancers occur on body parts exposed to the sun. They occur especially in fair-skinned people who spend considerable time outdoors. Australia has the highest rate of skin cancer in the world: At current rates of incidence, half of all Australians will develop some kind of skin cancer during their lifetimes. The most deadly type, malignant melanoma, is the most common cancer among Australians aged 15 to 44 years.[4] Scientists estimate that for every 1 percent decrease in the ozone layer, there will be an increase of 2 percent in UVB radiation at the earth's surface, and an increase of 3 to 6 percent in the incidence of skin cancer.[5]

UVB radiation puts the human skin in double jeopardy. Not only can it induce the growth of a cancer, but it can also suppress the immune system's ability to fight cancer as well as herpes and other infectious diseases.

Besides the skin, the other part of the body most exposed to light is the eye. UVB light can burn the cornea, causing a painful condition known as "snow blindness" because it afflicts skiers and mountaineers at high altitudes. Occasional snow blindness is very painful; repeated snow blindness

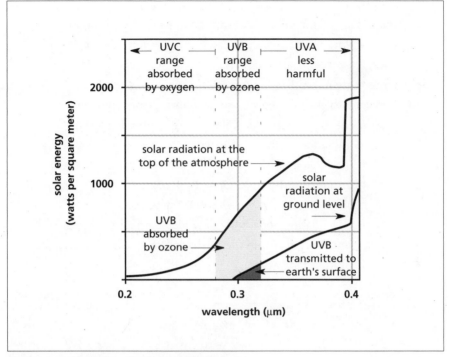

FIGURE 5-2 Absorption of Light by the Atmosphere
Incoming ultraviolet light from the sun is almost totally absorbed by oxygen and ozone in the atmosphere. Ozone particularly absorbs radiation in the range called UVB, which is dangerous to living things. (Source: UNEP.)

can reduce vision permanently. UVB light can also damage the retina and generate cataracts in the lens of the eye.

If more UVB light reached the earth's surface, any animal with eyes and skin exposed to the sun would be expected to suffer effects similar to those in human beings. Detailed studies of other effects of UVB are just beginning, but some results are already clear:

- Single-celled and very small organisms are more likely to be damaged than large organisms, because UVB can penetrate only a few layers of cells.
- UVB light penetrates only the top few meters of the ocean, but this is the layer where most aquatic microorganisms live. Research shows that these small, floating plants and animals are sensitive to UVB radiation.[6] There is not yet consensus on the

magnitude of the effect or about the impact of UVB on interactions among the various species in an ecosystem. But these microorganisms are the base of most ocean food chains. Therefore an increase in UVB could perturb many ocean species.

- Exposure to UVB light decreases leaf area, plant height, and photosynthesis in green plants. Different agricultural crops respond to UVB radiation to different extents, but for 60 percent of the crop plants studied, yields go down as UVB goes up. For example, one study suggested that a 25 percent depletion in the ozone layer could reduce soybean yields by 20 percent.[7]

- Ultraviolet radiation apparently degrades outdoor polymers and plastics, and it is a factor in the formation of low-level ozone, an ingredient in urban smog.

Living creatures have evolved to protect themselves from ultraviolet light in many ways—pigmentation, hair or scale coverings, mechanisms to repair damaged DNA, and behavior patterns that keep sensitive creatures hidden from strong sunshine. Because these devices work better for some species than others, one effect of a degraded ozone layer would be population declines or extinctions in some species and population increases in others. Grazers could grow out of balance with their forage supply, pests with their predators, parasites with their hosts. Every ecosystem would feel the effect of a diminished ozone layer in ways that are impossible to predict, especially if other changes, such as global warming, are going on at the same time.

The First Signals

In 1974 two scientific papers were published independently, both of which suggested a threat to the ozone layer. One said that chlorine atoms in the stratosphere could be powerful ozone destroyers.[8] The second said that CFCs were reaching the stratosphere and breaking up, releasing chlorine atoms.[9] Taken together, these publications predicted that human CFC use could produce extremely serious consequences.

Because they are inert and insoluble, CFCs do not dissolve in rain or react with other gases. The wavelengths of sunlight that reach the lower atmosphere do not break their strong carbon–chlorine and carbon–fluorine bonds. About the only way a CFC molecule can be cleansed from the atmosphere is to rise high enough that it encounters short-wavelength ultraviolet light—the very light that never reaches the Earth's surface because ozone and oxygen filter it out. That radiation breaks up the CFC molecule, releasing free chlorine atoms.

That's where the trouble begins. Free chlorine (Cl) can react with ozone to make oxygen and chlorine oxide (ClO). Then the ClO reacts with an oxygen atom (O) to make O_2 and Cl *again*. The Cl atom can then turn another ozone molecule into oxygen and be regenerated yet again (figure 5-3).

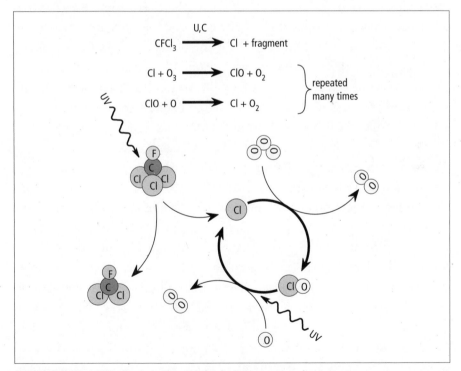

FIGURE 5-3 How CFCs Destroy Stratospheric Ozone
CFC molecules high in the stratosphere are broken up by ultraviolet light to release free chlorine atoms (Cl). These atoms react with ozone (O_3) to produce chlorine monoxide (ClO). The ClO then can react with an oxygen atom to release Cl again, which can react with another ozone molecule—and so on. The cyclic reaction is repeated over and over, greatly reducing the ozone concentration in the atmosphere.

A Cl atom can cycle through this series of reactions over and over, destroying one ozone molecule each time. The average Cl atom destroys about 100,000 ozone molecules before it is finally removed from action (by reacting with a substance such as methane or nitrogen dioxide that immobilizes it and brings it back down to earth).

The Delays

Overshoot requires delays, and there are many delays in the ozone system. The continuous regeneration of Cl means that many years will pass after its arrival in the stratosphere until a Cl atom finally ceases to break down ozone. Another is the long delay between the industrial synthesis of a CFC molecule and its arrival in the upper stratosphere. For some uses (such as aerosol propellants) production is followed quickly by discharge into the air. For other uses (such as refrigerants or foam insulation) the CFC is typically released into the air years after its production. After release, it takes decades for all CFC molecules to be cycled by the atmosphere's currents into the high stratosphere. Thus the thinning of the ozone layer measured at any time is a result of CFCs manufactured many years or decades earlier.

The processes that produce new knowledge and lead, eventually, to scientific consensus are also fraught with delay, though in this case several political factors reduced the time involved.

The two papers predicting the depletion of the ozone layer started a burst of research on atmospheric chlorine chemistry. In the United States, the scientific information also made its way quickly into the political process. That happened partly because the authors of one of the first papers were American, deeply worried about their findings, and energetic in bringing them to public attention (especially F. Sherwood Rowland, who took the matter to the U.S. National Academy of Sciences and the U.S. Congress). Another factor that created an impact in the U.S. was the well-organized environmental movement.

When American environmentalists understood the implications of the CFC–ozone connection, they went into action. They started by condemning the use of CFCs in aerosol spray cans. It was crazy, they said, to threaten life on earth just for the privilege of spraying on deodorant and shaving cream. The stigmatization of aerosol cans was oversimplified, since non-CFC aerosol propellants were also in use, and there were many other uses of CFCs. But aerosol cans were branded as ozone destroyers, and consumers responded: The sale of aerosol cans plummeted by more than 60 percent. You can see the results in figure 5-1; the growth temporarily stopped around 1975. Political pressure mounted for a law to ban the use of CFCs as aerosols.

There was industry resistance, of course. A DuPont executive testified before Congress in 1974 that "the chlorine–ozone hypothesis is at this time purely speculative with no concrete evidence to support it." But, he said, "If creditable scientific data . . . show that any chlorofluorocarbons cannot be used without a threat to health, DuPont will stop production of these compounds."[10] Not until 14 years later would DuPont, the world's largest producer of CFCs, honor that pledge.

A law forbidding the use of CFCs as aerosol propellants was passed in the United States in 1978. Together with the consumer action that had already reduced aerosol sales, that ban produced a large drop in worldwide manufacture of CFCs. In most of the rest of the world, however, aerosol sprays still contained CFCs, and other uses of CFCs, especially in the electronics industry, continued to climb. By 1980 world CFC use was back up to its 1975 peak and climbing (figure 5-1).

Overshoot: The Ozone Hole

In October 1984 scientists of the British Antarctic Survey measured a 40 percent decrease in ozone in the stratosphere over their survey site at Halley Bay in Antarctica. Their October ozone measurements had been declining steadily for about 10 years (figure 5-4). But the scientists had been reluctant to believe what they were seeing. A 40 percent drop seemed impossible. Computer models based on knowledge of atmospheric chemistry at the time were predicting only a few percent decline, at most.

The scientists rechecked their instruments. They looked for confirming measurements from somewhere else. Finally they found one: A measuring

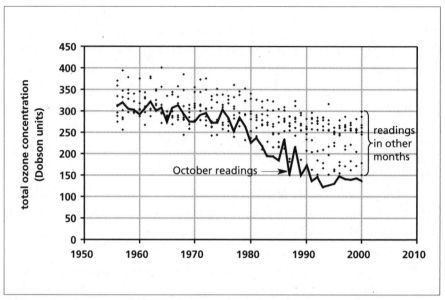

FIGURE 5-4 Ozone Measurements at Halley, Antarctica

Ozone concentrations in the atmosphere above Halley, Antarctica, measured during the month of October as the sun returns each southern spring, had been declining for more than a decade before the paper announcing the ozone hole was published in 1985. October ozone readings have continued to drop since then. (Source: J. D. Shanklin.)

station about 1,600 kilometers (1,000 miles) to the northwest also reported enormous declines in stratospheric ozone.

In May 1985 the historic paper was published that announced an "ozone hole" in the Southern Hemisphere.[11] The news shocked the scientific world. If true, the results proved that humankind had already exceeded a global limit. CFC use had grown above sustainable limits. Humans were already in the process of destroying their ozone shield.

Scientists at the National Aeronautics and Space Administration of the United States (NASA) scrambled to check readings on atmospheric ozone made by the Nimbus 7 satellite, measurements that had been taken routinely since 1978. Nimbus 7 had never indicated an ozone hole. Checking back, NASA scientists found that their computers had been *programmed to reject very low ozone readings* on the assumption that such low readings must indicate instrument error.[12]

Fortunately the measurements thrown out by the computer were recoverable. They confirmed the Halley Bay observations, showing that ozone

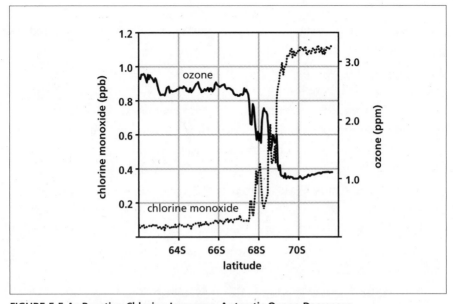

FIGURE 5-5 As Reactive Chlorine Increases, Antarctic Ozone Decreases
Instruments aboard NASA's ER-2 research airplane measured concentrations of chlorine monoxide and ozone simultaneously as the airplane flew from Punta Arenas, Chile (53°S) to 72°S. The data shown above were collected on September 16, 1987. As the plane entered the ozone hole, the concentration of chlorine monoxide increased to many times normal levels, while ozone levels plummeted. This finding helped establish the fact that chlorine-containing pollutants caused the ozone hole. (Source: J. G. Anderson et al.)

levels had been dropping over the South Pole for a decade. Furthermore, they provided a detailed map of the hole in the ozone layer. It was enormous, about the size of the continental United States, and it had been getting larger and deeper every year.

Why a hole? Why over Antarctica? What did this finding portend for the entire planet's protection from UVB radiation? The work of scientists over the next few years to solve this mystery was extraordinary. One of the most spectacular pieces of evidence that chlorine was indeed the culprit was gathered in September 1987, when scientists flew an airplane from South America directly toward the South Pole and into the ozone hole. Their measurements of ozone and ClO as they flew are shown in figure 5-5. Rises and drops in ozone are almost exactly mirrored by drops and rises in ClO.[13] Furthermore, the measured ClO concentrations in the "hole" are hundreds of times higher than any level that could be explained by normal atmospheric chemistry. This figure is often referred to as the "smoking gun" that proved, even to CFC manufacturers, that the ozone hole is no normal phenomenon. It is a sign of a highly perturbed atmosphere, caused by human-produced chlorine-containing pollutants.

It took several years for scientists to come up with an explanation for the hole. In a nutshell, here it is.

Since Antarctica is surrounded by oceans, winds can circle around the continent uninterrupted by landmasses. In the southern winter they create a *circumpolar vortex*, a whirl that traps air over Antarctica and keeps it from mixing with the rest of the atmosphere. The vortex creates a "reaction vessel" of polar atmospheric chemicals. (There is not such a strong vortex around the North Pole, so the northern ozone hole is less pronounced).

In winter the Antarctic stratosphere is the coldest place on Earth (down to −90 degrees C). In that extreme cold, water vapor hovers as a fog of minute ice crystals. The ice serves as a catalyst; the surfaces of these innumerable crystals enhance the chemical reactions that break up CFCs to release ozone-destroying chlorine.

The chlorine atoms formed in the dark of the Antarctic winter do not immediately enter the chain reaction of ozone destruction. Instead, each chlorine atom reacts just once with ozone to form ClO. Two ClO molecules come together to form a relatively stable ClOOCl dimer. An accumulation of these dimers builds up, poised for the return of the sun.[14]

Every September or October, in the Antarctic spring, the solar radiation breaks up the ClOOCl molecules to release an enormous burst of Cl, which goes to work on the ozone. Ozone concentration plummets.

Gradually the returning sunlight dissipates the circumpolar vortex, allowing south polar air to mix again. Ozone-depleted air is dispersed over the rest of the globe, while ozone levels over Antarctica return nearly to normal.

Lesser holes have been observed over the North Pole in the northern spring. Discrete holes are not expected to be found elsewhere, but as the gases in the atmosphere mix, the concentration of ozone in the stratosphere above the whole earth decreases. Because of the long lifetimes of CFCs and Cl in the atmosphere, the depletion will last for a long time—at least a century. Thus, once humanity moved beyond the limit (defined by the maximum sustainable rate of emission of CFCs), it destined itself to a long period of less-than-normal ozone protection against UVB—even if emissions were stopped immediately. Overshoot was, and will be for a long time, a reality.

The Next Response: Delays in Practice

There is some disagreement among the people who were involved in the global negotiations about whether the announcement of the ozone hole in 1985 energized politicians as thoroughly as it did scientists. International discussions were already under way to limit CFC production, but they had not made much progress. A meeting in Vienna, held two months before the ozone hole was announced, produced a feel-good statement that nations should take "appropriate measures" to protect the ozone layer, but it set no timetables and stipulated no sanctions. Industry had abandoned its search for CFC substitutes, since it was not apparent that they would be needed any time soon.[15] The Antarctic ozone hole would not be definitely linked to CFCs until three years later.

Something did happen politically between March 1985 in Vienna, when there was no real action, and September 1987 in Montreal, when the first international ozone-layer protocol was signed by representatives of 47 nations. The hole over the Antarctic certainly had a psychological effect, maybe all the more so because it was not understood. There was no doubt that the ozone layer was doing strange things. Though there was as yet no proof, there was enough science to point to CFCs as the likely culprits.

Proof or no proof, probably nothing would have happened had it not been for the United Nations Environment Program (UNEP), which hosted and prodded the international political process. Its staff assembled and interpreted the scientific evidence, presented it to governments, created a neutral venue for high-level discussions, and acted as mediators. UNEP's director, Mustafa Tolba, proved a skilled environmental diplomat, remaining neutral in the many squabbles that arose, patiently reminding everyone that no short-term selfish consideration was as important as the integrity of the earth's ozone layer.

The negotiating process was far from easy.[16] The world's governments were confronting a global environmental problem before it was completely understood and before it had produced any measurable damage to human health or the economy. Major CFC-producing nations played predictable roles in trying to block cutbacks in CFC use. Critical decisions sometimes hung on delicate political threads. The United States, for example, played a strong leadership role, which was several times nearly undercut by deep divisions within the Reagan administration. Those divisions came to public attention when Interior Secretary Donald Hodel said in public that the ozone layer would be no problem if people would just wear broad-brimmed hats and sunglasses when they went outside. The international ridicule heaped upon that statement (including cartoons of cows and dogs and trees and corn plants wearing hats and sunglasses) helped those members of the administration who were trying to get the president to take the ozone problem seriously.

Meanwhile, UNEP pressed on. As environmental groups in Europe and the United States put heat on their governments, scientists conducted workshops to educate journalists, parliamentarians, and the public. Responding to pressures from all sides, national governments finally—and surprisingly quickly—signed in Montreal in 1987 a Protocol on Substances That Deplete the Ozone Layer.

The Montreal Protocol stipulated first that world production of the five most commonly used CFCs should be frozen at 1986 levels. Then production should be reduced by 20 percent by 1993, and by another 30 percent by 1998. This "freeze-20-30" agreement was signed by all the major producers of CFCs.

The Montreal Protocol was historic, going far beyond what environmentalists at the time thought was politically possible. But it soon became apparent that the CFC reductions it called for were not enough. Figure 5-6 shows what would have happened to the concentration of ozone-destroying Cl in the stratosphere if emissions were cut according to the Montreal Protocol (and also to each of the subsequent agreements, in London, Copenhagen, Vienna, and Montreal again; more on this below). Despite the production cuts, the large stocks of CFCs that had been produced but not yet released, and that had been released but had not yet reached the stratosphere, would have continued to raise the Cl concentration.

The reasons for the weakness of the agreement were understandable. Most industrializing countries did not sign. China, for example, was trying to equip millions of households with their first refrigerators, which meant a huge new demand for Freon. The USSR waffled, saying that its five-year planning process did not allow rapid change in CFC production. It

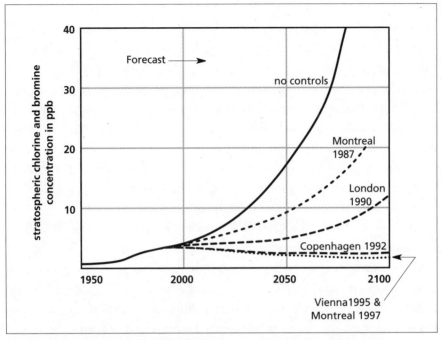

FIGURE 5-6 Projected Growth of Stratospheric Inorganic Chlorine and Bromine Concentrations Due to CFC Emissions
Past and projected stratospheric abundances of chlorine and bromine under various policies: no protocol, under the original Montreal Protocol's provisions, and under later additional agreements. Sustaining the 1986 production rate of CFCs would have led stratospheric chlorine concentrations to increase by a factor of 8 by 2050. The first Montreal Protocol defined lower emission rates, but it would still have permitted chlorine levels to climb exponentially. The London Agreement phased out most but not all CFC use; it still would have resulted in increasing chlorine levels beginning around the year 2050. Subsequent agreements have slowly tightened the restrictions on uses of chlorine-releasing chemicals, leading to projections of decreasing chlorine levels in the stratosphere after the year 2000. (Sources: WMO; EPA; R.E. Bendick.)

demanded and got a slower phase-down schedule. Most industrial makers of CFCs were still hoping to maintain at least part of their market.

Within a year after the Montreal Protocol was signed, however, even greater ozone depletions were measured, and the "smoking gun" evidence was published. At that point DuPont announced that it would phase out its manufacture of CFCs completely. In 1989 the United States and the European Union declared that they would stop all production of the five most common CFCs by the year 2000. They called upon the world to invoke the stipulation that had been written into the Montreal document requiring periodic reassessment of the ozone situation and stronger measures, if necessary.

After further negotiations, again led by UNEP, governments from 92 countries met in London in 1990 and agreed to phase out all CFC production by the year 2000. They added to the phaseout list methyl chloroform, carbon tetrachloride, and halons, which are also ozone-destroying chemicals. Several industrializing countries refused to sign unless an international fund was established to help them with the technical shift to CFC alternatives. When the United States balked at contributing to that fund, the agreement almost failed, but in the end the fund was established. The evolution in stratospheric chlorine (and bromine, another ozone-depleting chemical) expected from the London Agreement is shown in figure 5-6.

In spring 1991 new satellite measurements over the Northern Hemisphere showed ozone depletion occurring about twice as fast as expected. Over populated areas in North America, Europe, and Central Asia, depressed ozone levels extended for the first time into summer, when radiation damage is likely to harm both people and crops. Later in the 1990s below-average ozone levels were reported as far south as Spain.

As a consequence of this alarming news, many countries, headed by Germany, moved to phase out CFC and halon production even faster than the London Agreement required. Many multinational corporations, especially in the electronics and automobile industries, did the same. Some developing countries, such as Mexico, announced they would not take advantage of their 10-year grace period, but would follow the same reduction schedule as the industrialized countries. Gradually all others, including China and India, followed suit, and currently all production is scheduled to cease by 2010.

At yet another negotiating session, in Copenhagen in 1992, the signers of the Montreal Protocol agreed to advance the phaseout once more—to eliminate new production of halons by 1994 and all CFCs by 1996 and to put a cap on emissions of the soil fumigant methyl bromide, a powerful ozone depleter that had not even been discussed in London. According to atmospheric models available at that time, it was believed the Copenhagen "tightening" would bring the ozone layer back to its 1980 level 10 years sooner than would the London measures (by the year 2045 instead of 2055). It would reduce cumulative ozone loss by 28 percent, preventing 4.5 million cases of skin cancer and 350,000 cases of blindness.[17] Later it became clear that the Copenhagen "tightening" was, in fact, necessary to achieve any reduction at all in the chlorine/bromine concentration (see figure 5-6).

By 1996, 157 nations had become parties to this strengthened agreement. There was little more that could be done. Minor adjustments were made to the agreement in Montreal in 1997, on the 10th anniversary of the first protocol. The Scientific Assessment of the Ozone Layer—1998 (conducted under the aegis of the World Meteorological Organization and UNEP)[18] noted that "based on past emissions of ozone-depleting substances and a projection of maximum allowances under the Montreal Protocol into the future, the maximum ozone depletion is estimated to lie within the current decade or the next two decades." Four years later, in the 2002 Scientific Assessment: "Antarctic ozone levels will be increasing by 2010. A return to pre-1980 ozone amounts is expected in the middle of this century."[19] It was too late to affect the peak years of ozone depletion—from 1995 to 2010—because the chemicals that would induce them were already wafting slowly up to the stratosphere. To be sure that the peak was indeed a peak and that the ozone layer would eventually recover, the main task was to carry out and enforce the protocol. The Conference of Parties to the Montreal Protocol continued to meet and improve on the agreement: For example, in 1999 in Beijing, participants agreed to increase the multilateral fund that helps finance efforts of developing countries to meet their deadlines. Currently, other substances are being added to the regimen, and trade in ozone-depleting substances is being banned.

By 2000 world production of "CFC gases" had fallen from its 1988 peak of more than one million tons to less than 100,000 tons per year (figure 5-1).[20] Industry had adjusted to the phaseout of these important chemicals with much less expense and disruption than anyone would have guessed when the international negotiations began. (The final price tag, including the cost of negotiations and enforcement, is estimated at $40 billion.[21]) Since CFCs are also greenhouse gases, several thousand times as powerful as carbon dioxide, their phaseout will also help reduce the rate of global climate change. A less damaging substitute, called HCFCs, is still produced at a rate of half a million tons per year (figure 5-1).

All the while, news kept coming in from the stratosphere—in bits and pieces. In 1995 and 1996 ozone concentration over the North Pole reached new historical lows, including one brief 45 percent drop over Siberia. Northern midlatitude ozone layer losses in winter and spring 1998 averaged 6 to 7 percent. In fall 1998 the ozone hole over the South Pole was larger and deeper than any that had yet been detected[22]—until the same claims were made in 2000 and then in 2003. And although the growth in the ozone hole became progressively slower, in 2002 the WMO Scientific Assessment could not yet "say whether the area of the (Antarctic) ozone hole had maximized"—although they did agree that "the ozone layer will slowly recover over the next 50 years."[23]

The ozone layer will remain in its most vulnerable state over the first two decades of the twenty-first century. If the Montreal Protocol and subsequent agreements are honored, if illegal manufacture stops, and if there are no major volcanic eruptions (which can also deplete stratospheric ozone for short periods), the ozone layer should return approximately to its original state by about the year 2050.

One hitch in this happy story is the rise of CFC smuggling. Though the governments of the United States and Europe forbade both the manufacture and the import of newly made CFCs, many of their citizens were willing to pay a high price for recharging their car air conditioners or their cooling units. In the United States a high excise tax on new CFCs, designed to encourage recycling, pushed the price even higher. Countries permitted under the treaties to go on making CFCs until 2010 (primarily Russia, China, and India) found it hard to resist such a lucrative market. Smugglers use tricks such as falsely labeling new CFCs as recycled. The U.S. Justice Department reported that the profits on illegally imported CFCs were higher than those on cocaine. It is impossible to get exact figures on the extent of the market in illegally imported CFCs—estimates range from 20,000 to 30,000 tons per year[24]—but it has not been large enough to interrupt the downward trend in total CFC production.

Despite this and other minor issues, the world has largely reached consensus on the problem and has made enormous progress in implementing solutions. It has taken more than 25 years, but successful response to overshoot is clearly possible.

Getting Along without CFCs

While the diplomacy was going on, industrial creativity was coming up with ways to reduce the release of existing CFCs and employ substitutes. One-third of the problem has been solved through conservation, simply reducing the need for the chemicals. For example, increased insulation can reduce the requirements for refrigeration. And recycling the chemicals for reuse can reduce the emissions. One-third of the solution relies on the interim use of substitutes, such as hydrogenated CFCs, called HCFCs, which are only 2 to 10 percent as destructive to the ozone layer. Scheduled for phasing out by 2030; their use gives time to find more permanent solutions. The final third has been accomplished by shifting to alternatives that do not harm the ozone layer at all.

Because of the 1978 ban in the United States, manufacturers had already adopted other kinds of aerosol propellants, most of which proved less expensive than CFCs. As atmospheric chemist Mario J. Molina said: "In 1978, when the United States prohibited the use of CFCs as propellants in spray cans, experts said the ban would put a lot of people out of work. It didn't."[25]

Coolants in refrigerators and air conditioners used to be released into the air when those units were serviced or discarded. Now recycling devices capture, purify, and reuse those coolants. In the United States CFC recycling—and also the repair of leaks—is encouraged by a substantial tax, which makes recycling profitable. A current challenge is to keep the safe replacement substances apart from their ozone-depleting predecessors in the recycling process.

Electronics and aeronautics firms worked out substitute solvents for cleaning circuit boards and airplane parts, some of them involving simple water solutions. They also reworked manufacturing processes to eliminate washing steps entirely, with considerable economic savings. Firms from the United States and Japan formed a coalition to share without charge their research on these adaptations with electronics manufacturers all over the world.[26]

Chemical companies started marketing hydrogenated CFCs and other new compounds to substitute for specific uses of CFCs. Car air conditioners now contain a CFC substitute called HFC-134a. The additional cost of this new coolant is not the predicted $1,000 to $1,500 per car, but more like $50 to $150.

Insulating plastic foam is now injected with other gases; hamburgers are wrapped in paper or cardboard, not CFC-containing plastic; environmentally conscious consumers use washable ceramic coffee cups instead of throwaway plastic ones.

Growers of cut flowers in Colombia discovered they could practice integrated pest management instead of sterilizing soil with methyl bromide. Kenyan farmers started using carbon dioxide instead of methyl bromide to fumigate stored grains. Zimbabwean tobacco growers tried crop rotation instead of methyl bromide. A UNEP study concluded that 90 percent of methyl bromide use could be replaced with other pest control measures, often at reduced cost.

The Moral of the Story

A report by 350 scientists from 35 nations, coordinated by the World Meteorological Organization in 1999 gives the consensus view about prospects for the ozone layer.

> The ozone depletion caused by human-produced chlorine and bromine compounds is expected to gradually disappear by about the middle of the twenty-first century as these compounds are slowly removed from the stratosphere by natural processes. This environmental achievement is due to the landmark international agreement to control the production and use of ozone-depleting substances.[27]

There are many lessons to be drawn from the ozone story, depending upon your worldview and political predilections. Here are the lessons we see:

- Frequent monitoring of important attributes of the environment is essential, as is the quick and honest reporting of results.
- Political will can be summoned on an international scale to keep human activities within the limits of the Earth.

- International agreement to avoid future damage to the environment typically requires both the tools and the will to make long-term projections.
- People and nations do not have to become perfect saints in order to forge effective international cooperation on difficult issues; nor is perfect knowledge or scientific proof necessary for action.
- A world government is not needed to deal with global problems, but it is necessary to have global scientific cooperation, a global information system, an international forum within which specific agreements can be worked out, and international cooperation in enforcing those agreements.
- Scientists, technologists, politicians, corporations, and consumers can react quickly when they see the need to do so—but not instantly.
- Dire predictions of industry about the economic consequences of meeting environmental regulations may be exaggerated. This can be the result of deliberate distortions designed to slow political change. More likely it comes from systematically underestimating the capacity for technological advance and social change.
- When knowledge is incomplete, environmental agreements need to be written flexibly and reviewed regularly. Constant monitoring is needed to keep track of the problem, to make adjustments if necessary, and to chart improvements. Never assume a global problem has been forever solved.
- All the actors in the ozone agreement were important and will be needed again: an international negotiator like UNEP; a few national governments willing to take the political lead; flexible and responsible corporations; scientists who can and will communicate with policy makers; environmental activists to apply pressure; consumers willing to shift product choices on the basis of environmental information; and technical experts to come up with innovations that can make life possible, convenient, and profitable even when it must adapt to bring human impacts back within limits.
- Of course, we can also see in the ozone story all the ingredients of an overshoot and collapse system—exponential growth, an

erodable environmental limit, and long response delays, both physical and political. It took 13 years from the first scientific warnings in 1974 to the signing of the Montreal Protocol in 1987, and 13 more years from the first signing until the strengthened protocol was due to be fully implemented in 2000. It may take longer than that to round up the remaining noncooperators, cheats, and smugglers. It will take more than a century for the chlorine to be fully cleansed from the stratosphere after 2050.

This is a story of overshoot. And a story of how humanity is easing back toward sustainable behavior. Everyone hopes it will not be a story of collapse. Whether it will be or not depends on how reversible the damage to the ozone layer is and whether future atmospheric surprises will appear. It also depends on the ability to remain vigilant and block efforts by special-interest groups, and their politicians, to secure exemptions from the ban on ozone-destroying chemicals. If those conditions are met, the rise and fall of the stratospheric ozone hole, can serve as an inspiration in the efforts to confront other global limits.

CHAPTER 6

Technology, Markets, and Overshoot

All the evidence suggests that we have consistently exaggerated the contributions of technological genius and underestimated the contributions of natural resources. . . . We need . . . something we lost in our haste to remake the world: a sense of limits, an awareness of the importance of earth's resources.

—STEWART UDALL, 1980

Homo sapiens has lived on Earth for 100,000 years. Humans have cultivated the land and organized themselves into cities for about 10,000 years. They have experienced rapid exponential growth of population and capital for about 300 years. During those last few centuries, spectacular technical and institutional innovations—the steam engine, the computer, the corporation, international trade agreements, and many other changes— have allowed the human economy to transcend apparent physical and managerial limits and keep on growing. Especially over the past few decades, the expanding industrial culture has instilled into nearly every community on earth the desire for and the expectation of ever-continuing material growth.

The idea that there might be limits to growth is for many people impossible to imagine. Limits are politically unmentionable and economically unthinkable. The culture tends to deny the possibility of limits by placing a profound faith in the powers of technology, the workings of a free market, and the growth of the economy as the solution to all problems, even the problems created by growth.

The most common criticisms of the original World3 model were that it underestimated the power of technology and that it did not represent adequately the adaptive resilience of the free market. It is true that we did not include in the original World3 model technological progress at rates that would automatically solve all problems associated with exponential growth

> **Twenty years ago some spoke of limits to growth. But today we now know that growth is the engine of change. Growth is the friend of the environment.**
>
> President George H. W. Bush, 1992

> **This is my long-term forecast in brief: The material conditions in life will continue to get better for most people, in most countries, most of the time, indefinitely. Within a century or two, all nations and most of humanity will be at or above today's Western living standards. I also speculate, however, that many people will continue to think and say that the conditions of life are getting worse.**
>
> Julian Simon, 1997

> **In 1972, the Club of Rome published "Limits to Growth," questioning the sustainability of economic and population growth. "Limits to Growth" estimated that by now we would begin to see declines in food production, population, energy availability and life expectancy. None of these developments has even begun to occur, nor is there any immediate prospect that they will. So the Club of Rome was wrong. . . .**
>
> ExxonMobile, 2002

in the human ecological footprint. That was because we did not—and still do not—believe such tremendous technological advance will occur by itself, nor through the unaided operation of "the market." Impressive—and even sufficient—technological advance is conceivable, but only as a consequence of determined societal decisions and willingness to follow up such decisions with action and money. Even with all that, the desired technology will only appear after significant time delays. This is our perspective on reality today, as it was our view 30 years ago. This view is reflected in World3.[1]

Technological advance and the market are reflected in the model in many ways. We assume in World3 that markets function to allocate limited investment capital among competing needs, essentially without delay.[2] Some technical improvements are built into the model, such as birth control, resource substitution, and the green revolution in agriculture. In several scenarios we test accelerated technological advance and possible future technical leaps beyond these "normal" improvements. What if materials are almost entirely recycled? What if land yield doubles again and yet again? What if emissions are reduced at 4 percent per year over the coming century?

Even with such assumptions, the model world tends to overshoot its limits. Even with the most effective technologies and the greatest economic resilience that we believe is possible, *if those are the only changes*, the model tends to generate scenarios of collapse.

In this chapter we will explain why. Before we go on, however, we need to acknowledge that we are dealing here with issues that are not only subjects of scientific study but also articles of faith. If we suggest that technology or markets have problems or limits, some will consider us to be heretics, and they will say that we are anti-technology.

This is simply not true. Donella received her doctorate from Harvard University; Dennis and Jorgen both graduated with PhDs from the Massachusetts Institute of Technology—both institutions are leading developers of new technology. We all have had a profound respect and a great enthusiasm for the powers of science to solve humanity's problems. One small reminder to us of the wonderful power of technical advance is provided by our work on this series of books. In 1971 we wrote *The Limits to Growth* on electric typewriters, we drew graphs by hand, and we needed a huge mainframe computer to run World3. It required 10 to 15 minutes to generate a single scenario. In 1991 we revised the model, wrote a new book, prepared graphs and charts, and laid out pages on desktop computers. Producing a scenario of World3 over 200 simulated years required three to five minutes. In 2002 we could run the World3 model on laptops, collaborate in revising the book via the Internet, and store all our results on a CD-ROM. Now it takes about four seconds to run the model. We count on technical efficiencies to help ease the human ecological footprint back down below the earth's limits with elegance and with minimal sacrifice.

Nor are we anti-market. We understand and respect the capacities of the market. Two of us have PhDs from a major business school; Jorgen was president of the Norwegian School of Management for 8 years. Dennis was a faculty member at Dartmouth's Tuck School of Business for 16 years. We participate in the management boards of high-technology companies. We have all experienced firsthand the difficulties and absurdities of centrally planned economies. We count on improvements in market signals, as well as in technologies, to bring about a productive and prosperous sustainable society. But we do not have faith, and we have no objective basis for expecting, that technological advance or markets, by themselves, unchanged, unguided by understanding, respect, or commitment to sustainability, can create a sustainable society.

Our qualified faith in technology and markets is based on our understanding of systems. It comes from the discipline of having to express in nonlinear, feedback-based models exactly *what technology is*, and *what markets do*. When one has to model these systems concretely, instead of making sweeping general claims for them, one discovers their functions and powers in the economic system and also their limitations.

In this chapter we will:

- Describe technology and market feedback processes as we understand them and have modeled them in World3.
- Show computer runs in which we assume more and more effective technologies to overcome limits.
- Explain why overshoot and collapse are still the dominant modes of behavior in these runs.
- End with two short case studies, one about oil, one about fisheries, which demonstrate how technologies and markets in the present world do not guarantee a smooth transition to sustainability.

Technology and Markets in the "Real World"

What "really" is technology? Is it the ability to solve any problem, the physical manifestation of the inventive genius of humankind? Is it a steady exponential increase in the amount that can be produced by an hour of labor or a unit of capital, mastery over nature? Is it the control of some people by other people using nature as their instrument?[3] Human mental models contain all these concepts of technology, and more.

What "really" is the market? Some would say it is simply the place where buyers and sellers come together to establish exchange prices that express the relative values of commodities. Others would say the free market is a fiction invented by economists. Some who never experienced a market that was free of bureaucratic control came to think of it as a magical institution that somehow delivers consumer goods in abundance. Is the market the right and ability to own capital privately and keep the returns? Or is it the most efficient means of allocating society's products? Or is it a device by which some people control other people using money as their instrument?

We believe the following processes are in the model that people most commonly have in mind when they say that technology and markets can eliminate the limits to growth:

- A problem related to limits appears: A resource becomes scarce, or a pollutant begins to build up.
- The market causes the price of the scarce resource to rise relative to other resources, or the pollutant begins to exact costs that are reflected in rising prices of products or services that generate that pollutant. (Here there is usually an admission that the market needs significant correction in order to reflect the cost of "externalities" like pollution.)
- The rising price generates responses. It pays geologists to go find more of the resource, or biologists to breed more, or chemists to synthesize it. It causes manufacturers to substitute a more abundant resource for the scarce one and to rely more on recycling. It forces consumers to use fewer products containing the resource or to use the resource more efficiently. It induces engineers to develop pollution control devices, or to find places to sequester the pollutant, or to invent manufacturing processes that don't produce the pollutant in the first place.
- These responses on both the demand and the supply side compete in the market, where the interplay of buyers and sellers decides which technologies and consumption patterns solve the problem most quickly and efficiently at least cost.
- Eventually the problem is "solved." The system has overcome that particular scarcity or reduced the damage from that pollutant.
- All this is feasible at a cost society is willing to pay and occurs quickly enough to avoid irremediable damage.

This model does not rely solely on technology or solely on the market; it assumes there will be smooth, effective interactions between the two. The market is needed to signal the problem, to direct resources toward its solution, and to select and reward the best solution. The technology is needed to solve the problem. The whole package has to be working well. Without signals from the market, the technology will not be forthcoming. Without technical ingenuity, the market's signals will produce no result.

Notice also that this model takes the form of a negative feedback loop—a chain of causation that acts to reverse a change, correct a problem, restore a balance. The resource scarcity is overcome. The pollution is cleaned up or sequestered. The society can continue to grow.

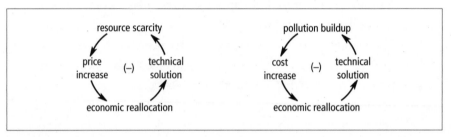

Negative Feedback Loops

We believe that adjustment loops like these do exist and are important. We have included them in many places in World3, but not as a single, aggregate, wonder-working variable called "technology." Technologies arise in many places in the model and have many effects. Health care improvements, for example, are automatic in World3. They are generated and increase life expectancy whenever the simulated world's service sector can pay for them. Birth control technology appears in World3 when the health care system can support it and when a desire for smaller family size demands it. Improvements in land yield are also automatic in World3, as long as food demand is unsatisfied and capital is available.

If nonrenewable resources become scarce, the World3 economy allocates more capital to discovering and exploiting them. We assume that the initial nonrenewable resource base can be used completely, though as resources are depleted it takes more and more capital to find and extract those that remain. We also assume that nonrenewable resources are perfectly substitutable for each other without cost or delay. Therefore we lump them all together without distinguishing one from another.

By changing numbers in the model, we can strengthen or weaken these assumed market–technology adjustments. If we don't change the numbers, these technologies evolve in the simulated world at roughly the same stages of industrial output per capita at which they appeared in the present highly industrialized countries.

In World3 the need for the built-in technologies—health care, birth control, agricultural improvement, resource discovery and substitution—is signaled perfectly and without delay to the capital sector. Capital is allocated toward that need immediately as long as there is sufficient industrial or service output to make it possible. We do not represent prices explicitly, because we assume that prices are intermediary signals in an adjustment mechanism that works instantly and perfectly. We represent the mechanism ("scarcity produces a technical response") without the intermediary of a price. That assumption omits many delays and inaccuracies that occur in "real" market systems.

A number of other technologies in World3 do not become effective unless we turn them on in test scenarios. They include resource efficiency and recycling, pollution control, unconventional increases in land yield, and land erosion control. When we first built the model, we didn't consider these technologies so established that they were technically proven and readily adopted by anyone in the world who could pay for them.[4] We therefore programmed them so they could be activated as a discontinuous step at any simulated time that seemed reasonable to the model user. For instance, one could assume that the entire world would make a major commitment to recycling in 2005 or a concerted effort against pollution in 2015. In the current version of World3 these technologies are modeled as "adaptive technologies," which evolve gradually when there is need for more resources, less pollution, or more food in the simulated world.[5] It is left to the model user, however, to determine the strength of these technological responses. These "turn-on" technologies require capital, and they come on only after a development and implementation delay, which is normally set at 20 years.

One reason to have a computer model is to try out different assumptions, to explore different futures. We can, for example, look at Scenario 2, the last run we showed in chapter 4, where growth was ended by a pollution crisis, and we can ask: What if that simulated world responded to the rising curve of pollution by making an earlier, more determined investment in pollution control technology? Scenario 3, figure 6-1, shows what happens with that change.

Stretching the Limits with Technology in World3

In Scenario 3 and all further computer runs in this book, we continue to assume the larger amount of nonrenewable resources and the advancing extraction technologies that we used as the basis for Scenario 2. In concrete terms this means that we assume enough nonrenewable resources in the year 2000 to supply 150 years of consumption at that year's consumption rate. The resources are obtained at an annual cost of about 5 percent of society's industrial output. So Scenario 2 will be the basis of comparison for the technology and policy changes that follow.

We apply changes one at a time—first pollution control technology, then land yield technology, and so forth—not because we think the world is likely to apply just one technology at a time, but because that progression helps make the model's responses more understandable. In our own work with World3, even if we want to try three simultaneous changes, we apply them one at a time, so we can understand the effect of each one separately before trying to comprehend the combined, interacting effects of all at once.

For many economists technology is a single exponent in some variant of the Cobb-Douglas production function—it works automatically, without delay, at no cost, free of limits, and produces only desirable outcomes. No wonder economists are so rapturous about its potential to solve human problems! In the "real world," however, we cannot find technology with

FIGURE 6-1 Scenario 3: More Accessible Nonrenewable Resources and Pollution Control Technology

In this scenario we assume the same ample resource supply as in Scenario 2 as well as increasingly effective pollution control technology, which can reduce the amount of pollution generated per unit of output by up to 4 percent per year, starting in 2002. This allows much higher welfare for more people after 2040 because of fewer negative effects from pollution. But food production does ultimately decline, drawing capital from the industrial sector and triggering a collapse.

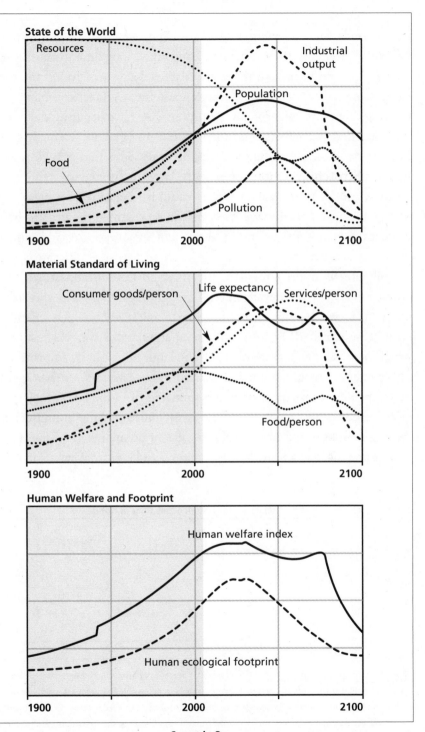

State of the World

Resources

Industrial output

Population

Food

Pollution

1900 · 2000 · 2100

Material Standard of Living

Consumer goods/person · Life expectancy · Services/person

Food/person

1900 · 2000 · 2100

Human Welfare and Footprint

Human welfare index

Human ecological footprint

1900 · 2000 · 2100

Scenario 3

those wonderful properties. The technologies we see are highly specific to particular problems; they cost money and take a long time to develop. Once they are proven in the lab there are further delays to develop the capital, labor, sales and service staff, marketing and finance mechanisms necessary to bring them into widespread use. Often they have negative, delayed, and unanticipated side effects. And the best technologies are jealously guarded by those who have patents on them, often disseminated at high price and with restrictive distribution agreements.

In World3 it is not possible, nor would it be useful, to represent technology in all its diversity. We have instead represented the process of technological advance for pollution abatement, resource use, and land yield with three summary parameters in each sector—the ultimate goal, the annual rate of improvement in the most successful laboratory, and the average delay between availability in the laboratory and widespread use in the field. As we describe each scenario, we will tell you which technologies have been activated. For the remaining simulations we will assume that when there is a need, technology in the laboratory can be improved at up to 4 percent annually. We assume that it takes 20 years, on average, for a new capability to be universally disseminated from the laboratory into the global production capital stock. Table 6-1 illustrates the consequences of these assumptions for emissions of persistent pollutants in Scenario 3.

Assume that a given stock of agricultural and industrial capital in the year

TABLE 6-1 Technology's Impact on Persistent Pollution Emissions in World3

Year	Percent Reduction
2000	0%
2020	10%
2040	48%
2060	75%
2080	89%
2100	95%

When technology can improve in the laboratory at 4 percent per year and be implemented throughout the global capital stock with an average delay of 20 years, it is possible to reduce emissions rapidly from their normal level. The table shows the percent reduction that are achieved in Scenario 3 of World3, after the model's population begins in the year 2002 to reduce pollution at the maximum rate permitted by technological advance.

2000 emits 1000 units of persistent pollution. If technology improves by 4 percent per year, and the dissemination delays average 20 years, only 900 units of persistent would be produced by the same capital stock in 2020. Emissions would fall almost half by 2040 and to only 5 percent of their original value by 2100. In World3, when the respective technologies are activated, similar gains are achieved for land yield and for efficiency of resource use.

In Scenario 3 we assume that in the simulated year 2002, before the global pollution level has risen high enough to cause great damage to either health or crops, the world decides to bring pollution down to the levels that prevailed in the mid-1970s and systematically allocates capital toward that goal. It chooses an "end of pipe" approach, abating pollution at the point of emission rather than reducing throughput at the source. Emissions are reduced as shown in table 6-1 with associated increases in capital investment costs of up to 20 percent. By the year 2100 the pollution level is brought down to the relatively low level that prevailed at the beginning of the twenty-first century.

In this scenario pollution continues to rise for nearly 50 years despite the abatement program, because of both delays in implementation and continued underlying growth in industrial production. But the pollution level stays much lower than it did in Scenario 2. It never gets high enough to affect human health, and hence this "global anti-pollution effort" succeeds in prolonging the era of high population and high welfare for essentially one more generation. The good times end in 2080, 40 years later than in Scenario 2, as measured by the human welfare index, which suddenly drops at that time. But pollution negatively affects land earlier in the century. Yields do not drop immediately, because the reduction in land fertility is partially offset by use of additional agricultural inputs. "Real world" examples of this phenomenon would be the use of lime to compensate for acid rain, the use of fertilizers to substitute for the lowered nutrient-generating capabilities of soil microbes poisoned by pesticides, and the use of irrigation to compensate for erratic rainfall caused by climate change.

The countertrends of declining soil fertility and increasing use of agricultural inputs in Scenario 3 lead to essentially stable food production from simulated year 2010 to 2030. The population goes on growing, however, so food per capita begins to turn down. But for several decades the output of the industry and service sectors remains sufficiently high to maintain acceptable living standards, despite the need for capital investment in agriculture and, later, for pollution control. In the last third of the twenty-first century, the pollution level has fallen so much that land fertility recovers. But the population pressure is large, and the amount of arable land declines due to urban sprawl and erosion. Furthermore, after the middle of the century industrial output falls rapidly, because so much capital has been pulled into the agricultural and pollution sectors that there is no longer enough investment to offset depreciation. The economy declines, and a collapse sets in, exacerbated late in the century by an increasing scarcity of nonrenewable resources.

The society in Scenario 3 greatly reduces its pollution level and succeeds in maintaining a high human welfare index for a long time. But eventually food becomes a problem. Scenario 3 could be described as a "food crisis." Of course in "real life" measures would be taken to sustain food availability at desired levels. What might happen if society turns its technological powers to the problem of raising more food? One possible result is shown in Scenario 4, figure 6-2.

In this run the pollution abatement program of Scenario 3 is activated again. At the same time, the model society decides in 2002 to respond aggressively to the stagnation in food per capita that emerged in the model system throughout the 1990s. Investment is shifted toward technologies intended to increase agricultural yield. The new technologies are assumed to take 20 years, on average, to implement in farmers' fields worldwide, and to raise yields by up to 4 percent per year, when there is a need. The investment in technology increases capital costs by 6 percent in 2040 and a full 8 percent in 2100. There is not much increase in yield up to 2050, because there is still enough food. But in the second half of the century average yields increase dramatically as a consequence of the exponential nature of the assumed technological advance.

The result is a long period of high population and high human welfare around the middle of the twenty-first century. The new agricultural technology helps raise food output from 2050 onward (compared with Scenario

FIGURE 6-2 Scenario 4: More Accessible Nonrenewable Resources, Pollution Control Technology, and Land Yield Enhancement
If the model world adds to its pollution control technology a set of technologies to increase greatly the food yield per unit of land, the high agricultural intensity speeds up land loss. The world's farmers end up trying to squeeze more and more food output from less and less land. This proves unsustainable.

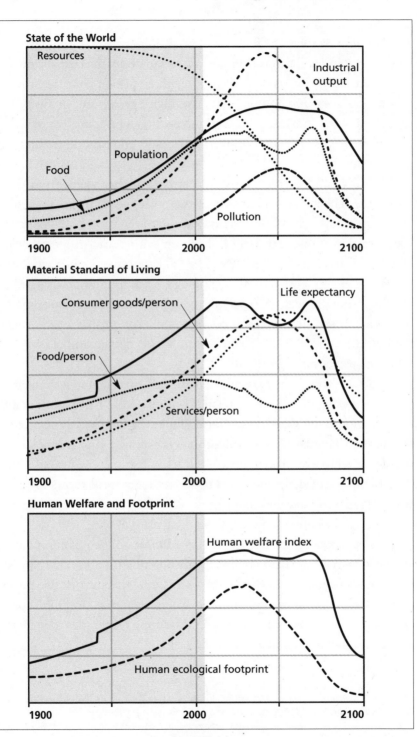

State of the World

Resources

Industrial output

Population

Food

Pollution

1900   2000   2100

Material Standard of Living

Consumer goods/person

Life expectancy

Food/person

Services/person

1900   2000   2100

Human Welfare and Footprint

Human welfare index

Human ecological footprint

1900   2000   2100

Scenario 4

3), but does not solve the food problem. Deterioration in land fertility and loss of arable land to erosion and to urban–industrial expansion finally overwhelm the positive effect of new technologies on yield, and total food production goes down after 2070. The high agricultural intensity in this simulated world induces galloping land erosion—not only loss of soils, but also nutrient loss, soil compaction, salting, and other processes that reduce land productivity.

With less land, farmers work to get even higher yields from the land remaining. The increased intensity of use causes yet more erosion, in a positive loop carrying the land system downhill. Scenario 4 could be termed a "land erosion crisis," reaching full bloom with a catastrophic fall in the amount of arable land after 2070. The fall cannot be counteracted in time by high-yielding agricultural technology, and food shortages trigger a decline in the population. The stressed agricultural sector pulls more and more capital and human resources from the economy, at a time when the diminishing nonrenewable resource base is also demanding capital. A fairly total collapse occurs before 2100.

Surely no sane society would pursue an agricultural technology that increases yields while destroying land. Unfortunately, there are examples of this behavior in the world today (for example, the land lost to salt accumulation in the Central Valley of California while nearby land is simultaneously being pushed to ever higher yields). But let us assume greater rationality on the part of coming generations. Let us add land protection technologies to the pollution control and yield-enhancing technologies. Scenario 5, figure 6-3, shows the results of all those changes taking place at once.

Here we assume, starting in 2002, in addition to the pollution reduction and yield-increasing technologies already described, a program that reduces global land erosion. As you remember, we assume that the first two programs require extra capital investment. But we assume that the third does not, because it mainly requires more careful farming techniques to increase the productive life of the soil.

FIGURE 6-3 Scenario 5: More Accessible Nonrenewable Resources, Pollution Control Technology, Land Yield Enhancement, and Land Erosion Protection
Now a technology of land preservation is added to the agricultural yield-enhancing and pollution-reducing measures already in place. The result is a slight postponement of the collapse at the end of the twenty-first century.

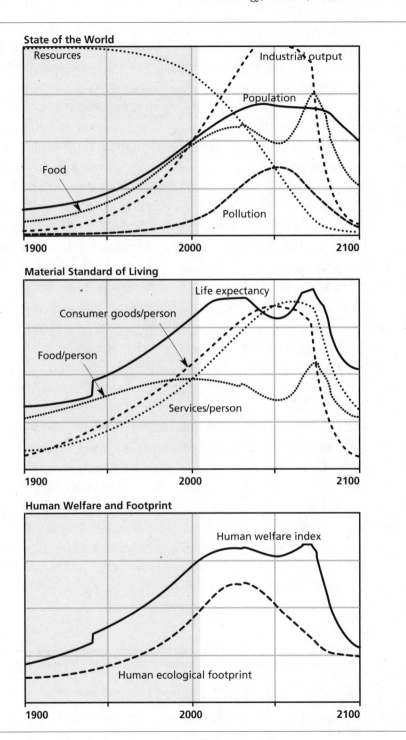

State of the World

Resources

Industrial output

Population

Food

Pollution

1900   2000   2100

Material Standard of Living

Life expectancy

Consumer goods/person

Food/person

Services/person

1900   2000   2100

Human Welfare and Footprint

Human welfare index

Human ecological footprint

1900   2000   2100

Scenario 5

This program does not have significant positive impacts until after 2050, when land erosion rates are reduced dramatically as a consequence of better farming techniques. The result permits a slight extension of the period of high human welfare after 2070. But the result is not sustainable. Scenario 5 ends in a collapse caused by the combination of crises in resources, food, and high costs more or less at once. Until about 2070 the average human welfare remains relatively high, despite unpleasant ups and downs in its various components. Food is roughly sufficient (though low during the middle third of the century), pollution is tolerable (though rather high during the middle third of the century), the economy grows (at least until 2050), services become more available, and life expectancy stays above 70 years. But after 2070 the costs of the various technologies, plus the rising costs of obtaining nonrenewable resources from increasingly depleted mines, demand more capital than the economy can provide. The result is rather abrupt decline. Scenario 5 could be described as a sum of many crises.

One might argue about which priority a society stressed in so many ways would drop first. Would it let land erode, let pollution rise, or get along with fewer raw materials? World3 assumes that materials and fuels would be given a high priority in order to go on producing the industrial output required to sustain investment in the other economic sectors. That particular choice, and the exact model behavior after investment capital becomes insufficient, is not important. We do not pretend to be able to predict what the world would do if it actually came to such a pass; we pay no attention to the model runs after the point when an important variable begins rapid decline. The important point is that such a predicament is possible, and it may well confront society.

If scarcity of nonrenewable resources is the final blow causing the collapse in Scenario 5, then a program of resource-saving technologies, added to all the others, should be able to help. Scenario 6, figure 6-4, shows the result.

FIGURE 6-4 Scenario 6: More Accessible Nonrenewable Resources, Pollution Control Technology, Land Yield Enhancement, Land Erosion Protection, and Resource Efficiency Technology

Now the simulated world is developing powerful technologies for pollution abatement, land yield enhancement, land protection, and conservation of nonrenewable resources all at once. All these technologies are assumed to involve costs and to take 20 years to be fully implemented. In combination they permit a fairly large and prosperous simulated world, until the bliss starts declining in response to the accumulated cost of the technologies.

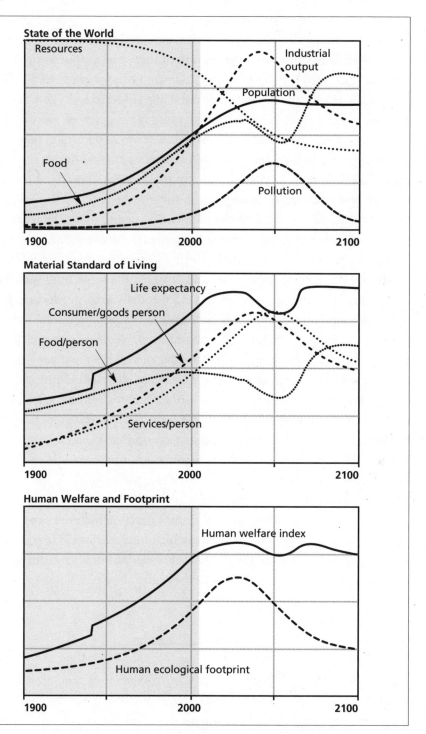

State of the World

Resources

Industrial output

Population

Food

Pollution

1900   2000   2100

Material Standard of Living

Life expectancy

Consumer/goods person

Food/person

Services/person

1900   2000   2100

Human Welfare and Footprint

Human welfare index

Human ecological footprint

1900   2000   2100

Scenario 6

We start in the simulated year 2002 a program to reduce the amount of nonrenewable resources needed per unit of industrial output by up to 4 percent per year. We keep the technological programs for improving pollution control, increasing land yield, and reducing land erosion. This, in short, amounts to a formidable twenty-first century program of increased eco-efficiency—with significant costs (capital costs are 20 percent higher in 2050 and 100 percent higher near 2090), but a significant reduction in the human ecological footprint as the goal.

This powerful combination of technologies helps avoid the collapse in Scenario 5 in the last third of the twenty-first century. But the technology program comes online just a bit too late to avoid a gradual decline in human welfare during the last one-third century. Population does not fall significantly, but life expectancy dips around 2050. At the same time—when pollution gets high enough to depress land fertility—food production is low; that effect is eventually overcome, however, by rising agricultural yields and pollution abatement technologies. Nonrenewable resources are depleted more slowly; their cost remains low. By the end of a rocky twenty-first century, a stable population of somewhat less than eight billion people is living in a high-tech, low-pollution world with a human welfare index roughly equal to that of the world of 2000. Life expectancy and food per capita are higher, service availability the same, but consumer goods per capita lower than at the start of the century. Industrial output begins to decline around 2040 because the rising expense of protecting the population from hunger, pollution, erosion, and resource shortage cuts into the capital available for growth. Service output per person and the level of material consumption begin to fall soon thereafter. Ultimately this simulated world fails to sustain its living standards as technology, social services, and new investment simultaneously become too expensive—a cost crisis.

Some Disclaimers

After a session of working with a model, computer or mental, it's a good idea to step back for a moment and remember that it is not the "real world" we have been experiencing, but a representation that is "realistic" in some respects, "unrealistic" in others. The task is to find insight in the model from those features of the scenarios that seem "realistic." It is also important to judge how the model's uncertainties or deliberate simplifications restrict its lessons. Following the preceding series of computer runs, we need to stop and regain perspective.

World3, we must remember, does not distinguish the rich parts of the world from the poor. All signals of hunger, resource scarcity, and pollution are assumed to come to the world as a whole and to elicit responses that draw on the coping capabilities of the world as a whole. That simplification makes the model very optimistic. In the "real world," if hunger is mainly in Africa, if pollution crises are mainly in Central Europe, if land degradation is mainly in the tropics, if the people who experience problems first are those with the least economic or technical capability to respond, there will be very long delays before problems are corrected. Therefore the "real" system may not respond as forcefully or successfully as does the World3 system.

The model's perfectly working market and smooth, successful technologies (with no surprising side effects) are also very optimistic. So is the assumption that political decisions are made without cost and without delay. We have to remember, too, that the World3 model has no military sector to drain capital and resources from the productive economy. It has no wars to kill people, destroy capital, waste land, or generate pollution. It has no ethnic strife, no strikes, no corruption, no floods, earthquakes, volcanic eruptions, nuclear accidents, AIDS epidemics, or surprising environmental failures. Therefore it is, in may ways, wildly optimistic. The model represents the uppermost possibilities for the "real world."

On the other hand, some people would say the technologies in the model are too limited. These critics would turn the technological cranks in the model much faster or even without limit (as in our Scenario 0). Our assumptions about discoverable resources, developable land, and absorbable pollution may be too low. Or they may be too high. We have tried to make them "realistic," given the data available to us and our own assessment of technical possibilities.

With all these uncertainties, we obviously should not study curves in the various scenarios with any assumption that they have quantitative precision. We don't take it as significant, for instance, that a food crisis appears in

Scenario 3 before a resource crisis does. It could easily happen the other way around. We are not predicting an industrial turndown starting in 2040, as in Scenario 6. The available numbers are just not good enough for World3, or any other model, to be interpreted that way.

So what, if anything, can we learn from these modeling exercises?

Why Technology and Markets Alone Can't Avoid Overshoot

The preceding tests can be summarized by saying that the human ecological footprint tends to grow beyond its sustainable level, and that this in turns triggers a forced decline in the footprint. Normally this decline is associated with a reduction in the average standard of living, through lower availability of food, fewer industrial or service goods per world inhabitant, or a higher pollution level in the human environment. The normal human reaction is to try to remove the constraint, in the hope of being able to continue the growth in population and economy.

One lesson from the six preceding runs is that in a complex, finite world, if you remove or raise one limit and go on growing, you encounter another limit. Especially if the growth is exponential, the next limit will show up surprisingly soon. There are *layers of limits*. World3 contains only a few. The "real world" contains many more, most of them distinct, specific, and local. Only a few limits, such as those associated with the ozone layer or the planet's climate, are truly global.

We would expect different parts of the "real world," if they keep on growing, to run into different limits in a different order at different times. But the experience of *successive and multiple limits* would unfold in any one place, we think, much the way it does in World3. And in an increasingly linked world economy, a society under stress anywhere sends out waves that are felt everywhere. Furthermore, globalization enhances the likelihood that those parts of the world involved in active trade with each other will reach many of their limits more or less simultaneously.

The preceding experiments also show that it is possible to reduce the human ecological footprint through development and use of technologies that reduce the materials and energy required by industry and agriculture.

When these technologies can be widely implemented, they permit a higher average standard of living within the same footprint. This is the much-touted dematerialization of the modern global economy.

A second lesson is that the more successfully society puts off its limits through economic and technical adaptations, the more likely it is to run into several of them at the same time. In most World3 runs, including many we have not shown here, the world system does not totally run out of land or food or resources or pollution absorption capability. What it runs out of is *the ability to cope*.

"The ability to cope" in World3 is represented, too simply, by the amount of industrial output available each year to be invested in solving problems. In the "real world" there are many other determinants of the ability to cope: the number of trained people; their motivation; the amount of political attention and intention; the financial risk that can be handled; the institutional capacity to develop, disseminate, and service new technologies; the managerial ability; the capacity of the media and political leaders to remain focused on crucial problems; the consensus among voters about important priorities; the degree to which people look far ahead to anticipate problems. All these capabilities can grow over time, if society invests in developing them. But at any one time, they are limited. They can process and handle just so much. When problems arise exponentially and in multiples, problems that could theoretically be dealt with one by one can overwhelm the ability to cope.

Time is in fact the ultimate limit in the World3 model—and, we believe, in the "real world." Given enough time, we believe humanity possesses nearly limitless problem-solving abilities. Growth, and especially exponential growth, is so insidious because it shortens the time for effective action. It loads stress on a system faster and faster, until coping mechanisms that have been adequate with slower rates of change finally begin to fail.

There are three other reasons why technology and market mechanisms that otherwise function well cannot solve the problems generated by a society driving toward interconnected limits at an exponential rate. They relate to goals, costs, and delays. The first reason is that markets and technologies are merely tools that serve the goals, the ethics, and the time horizons of the society as a whole. If a society's implicit goals are to exploit nature, enrich the elites, and ignore the long term, then that society will develop technologies and markets that destroy the environment, widen the gap between the rich and the poor, and optimize for short term gains. In short, that society develops technologies and markets that hasten a collapse instead of preventing it.

The second reason for the vulnerability of technology is that adjustment mechanisms have costs. The *costs* of technology and the market are reckoned in resources, energy, money, labor, and capital. These costs tend to rise nonlinearly as limits are approached. That fact is another source of surprising system behavior.

We have already shown in figures 3-19 and 4-7 how the wastes produced and the energy necessary to extract nonrenewable resources rise spectacularly as the resource grade declines. Figure 6-5 shows another rising cost curve: the marginal cost per ton of reducing nitrogen oxide emissions. It is fairly inexpensive to remove almost 50 percent of the emissions. There is a rising but still affordable cost for removing almost 80 percent. But then there is a limit, a threshold, beyond which costs of further removals rise enormously.

Further technical developments may well shift both curves to the right, making more complete cleanup affordable. Perhaps a different technology that eliminates the smoke completely could introduce another emission, connected to another abatement cost curve. Still, pollution abatement curves will always have the same basic shape. There are fundamental physical reasons why abatement costs soar as 100 percent abatement—that is, zero emissions—is demanded. A growing number of smokestacks or tailpipes guarantees those rising costs will be experienced. It may be affordable to cut pollutants per car in half, but if the number of cars then doubles, pollutants per car have to be cut in half again just to keep the same air quality. Two doublings will require 75 percent pollution abatement. Three doublings will require 87.5 percent.

Thus at some point it stops being true that growth will allow an economy to become rich enough to afford pollution abatement. In fact, growth takes an economy up a nonlinear cost curve to the point where further abatement becomes unaffordable. At that point a rational society would stop the expansion of its activity level, since further growth will no longer increase the welfare of its citizens.

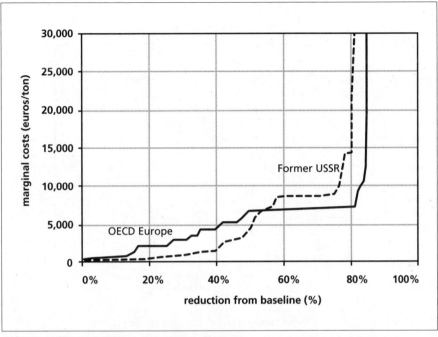

FIGURE 6-5 Nonlinear Costs of Pollution Abatement
The air pollutant NO_x can be removed from emissions to a significant degree at a low cost, but at some level of required abatement the cost of further removal rises precipitously. The marginal cost curve for NO_x removal is calculated for 2010 for OECD Europe and the former USSR in euros per ton. (Source: J. R. Alcamo et al.)

The third reason technology and the market can not automatically solve these problems is that they operate through feedback loops with information distortions and delays. *Delays* in market and technology responses can be much longer than economic theories or mental models expect. Technology–market feedback loops are themselves sources of overshoot, oscillation, and instability. One example of that instability, felt by all the world, was the fluctuation in oil prices during the decades after 1973.

An Example of Market Imperfection: Swings in the Oil Market

There were many causes of the "oil price shock" of 1973, but the most fundamental was the worldwide shortage of oil production capital (oil wells) relative to oil consumption capital (cars, furnaces, and other oil-burning

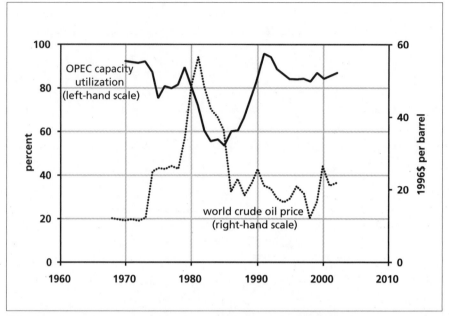

FIGURE 6-6 OPEC Oil Production Capacity Utilization and World Oil Price
With most of OPEC's production capacity in use in the 1970s, small interruptions in oil supply precipitated sudden and extreme price changes. The oscillations in oil prices took more than 10 years to unfold and caused economic turbulence all over the world, both on the way up and on the way down. (Source: EIA/DoE.)

machines). During the early 1970s the world's oil wells were working at over 90 percent capacity. Therefore a political upheaval in the Middle East that shut down even a small fraction of the world's oil production could not be offset by increasing supply elsewhere. This gave OPEC the opportunity to raise prices, and that is precisely what they did.

That price rise, and a second one for the same reason in 1979 (see figure 6-6), set off a wild set of economic and technical responses. On the supply side, more wells were drilled and more pumping capacity was installed outside the OPEC area. Marginal oil deposits suddenly became economical and were brought into production. Finding, building, and opening oil production facilities, from wells to refineries to tankers, took time.

Meanwhile, consumers were reacting to the higher prices by conserving. Car companies came out with more efficient models. People insulated their houses. Electric companies shut down their oil-burning generators and invested in coal-burning or nuclear facilities. Governments mandated various forms of energy saving and promoted the development of alternative energy sources. Those responses also took years. They ultimately resulted in long-lasting changes in the world's capital stock.

Advocates of the market seem to believe that it always acts quickly. But in the global oil market it was nearly 10 years before the many responses finally began to rebalance supply and demand—at the lower consumption rate consistent with the higher oil price. By 1983 global oil consumption had fallen by 12 percent compared with its peak in 1979.[6] Still there was too much oil production capital, and OPEC had to lower its pumping capacity even farther, down to nearly 50 percent. World oil price crept downward, and then plummeted in 1985, before it continued its downward trend (in real-dollar terms) to the end of the 1990s.

Just as the price had gone too far up, it then went too far down. As oil production facilities shut down and oil-producing areas were struck with depressions, conservation efforts were abandoned. Designs for more efficient cars were put on the shelf. Investment in alternative energy sources dried up. Ultimately, as these adjustment mechanisms gathered strength, they set up the conditions for the next imbalance and the next oil price rise, which may be what we are seeing in the relatively high oil prices in the early years after the turn of the millennium.

These overshoots and undershoots were a consequence of inevitable response delays in the oil market. They caused vast international shifts of wealth, enormous debts and surpluses, booms and busts and bank failures, all a result of trying to adjust the relative sizes of production capital and consumption capital for oil. None of these rises and falls in price was related to the actual underground quantity of oil (which was steadily going down) or to the environmental effects of drilling for, transporting, refining, and burning oil. The market's price signal mainly provided information about the relative scarcity or surplus of available oil.

For many reasons oil market signals have not yet given the world useful information about impending physical limits. The governments of producing nations intervene to raise the oil price; they have incentives to lie about their reserves, overestimating them in order to qualify for higher production quotas. Governments in the consuming nations work to keep prices low. They may lie about reserves, overestimating them in order to reduce the political power of individual producers. Speculators can amplify price swings. And the amounts of oil aboveground ready for use have much more influence on price than the amounts lying beneath the ground as future resources. The market is blind to the long term and pays no attention to ultimate sources and sinks, until they are nearly exhausted and it is too late for attractive solutions.

Economic signals and technological responses can evoke powerful responses, as the oil price example illustrates, but they simply aren't connected to the earth system in the right places to give society useful information about physical limits.

Finally, we want to return to the *purposes* to which technology and markets are put. They are simply tools. They have no more inherent wisdom or farsightedness or moderation or compassion than do the human bureaucracies that create them. The results they produce in the world depend upon who uses them and for what purposes. If they are used in pursuit of triviality, inequity, or violence, that is what they will produce. If they are asked to serve impossible goals, such as constant physical expansion on a finite planet, they will eventually fail. If they are called upon to serve feasible and sustainable goals, they can help bring about a sustainable society. In the next chapter we will show how that might occur.

Technology and markets, regulated and used toward the long-term communal good, will be of immense help. When the world decided to get along without CFCs, technology made that change possible over several decades. We don't believe it is possible to bring about a sufficient, equitable, sustainable world without technical creativity and entrepreneurship and a relatively free market. Neither do we believe they are enough. It will take other human abilities to make the human world sustainable. Absent those abilities, technological advance and markets can work in concert to thwart sustainability and to hasten collapse of important resources. That is precisely what happened to the ocean's fisheries.

Technology, Markets, and the Destruction of Fisheries

> I remember catching 5,000 pounds of fish in eight nets. Today it
> might take up to 80 nets. Back then, the average codfish in the spring
> would probably be 25 to 40 pounds. Now it's 5 to 8 pounds.
>
> —A FISHERMAN ON THE GEORGES BANK FISHING GROUND, 1988

> You want to know about cod, I'll tell you. There ain't no more.
>
> —CANADIAN FISHERMAN DAVE MOLLOY, 1997

The recent history of world fisheries is an illustration of how inadequately
the technology and markets may respond to approaching limits. In the case
of global fisheries, the "normal" combination of denial of limits, increased
effort to maintain traditional catch volumes, expulsion of foreign fishermen, subsidies to local fishermen, and finally hesitant societal regulation
of access came into play. In some cases—as in the cod fisheries on Canada's
eastern coast referred to in the quotes above—societal intervention came
too late to save the resources.

The move to regulation has gradually extended to most large fisheries.
The era of the "open seas" is certainly coming to an end. Limits are finally
obvious and are now a dominant aspect of world fisheries. As a consequence
of resource scarcity and regulation, the global catch of wild fish has stopped
growing. During the 1990s the total world commercial marine fish catch
oscillated below 80 million metric tons per year[7] (figure 6-7). There is no way
of knowing until many more years have passed whether this plateau is sustainable, or is the beginning of a collapse. Around 1990, the UN Food and
Agriculture Organization believed the world's seas could not sustain a commercial catch of more than 100 million metric tons per year from conventional resources—which is just above the level that we saw through the 1990s.

Not surprisingly, fish farming has grown rapidly during the same period,
now producing nearly 40 million tons per year, up from 13 in 1990. One-third
of the fish consumed in the world is now farmed. Shouldn't we be happy
with this response of the market and technology? Doesn't the growth of fish

FIGURE 6-7 World Catch of Wild Fish
The total world catch of wild fish increased dramatically from 1960 to 1990. But in the last decade of the twentieth century, total catch stopped growing. (Source: FAO.)

farming just illustrate the ability of technology and markets to solve the problem? Not really, and for three reasons. Fish production used to be a food source; it is becoming a food sink. Fish and other aquatic species used to feed the poor; they go now increasingly to the rich. Fish schools are a neutral part of the environment; fish farms are environmentally devastating.

First, ocean fisheries are a true source of food for humanity. They converted simple plants into delicious flesh. Fish farming is not a net food source; it merely converts one form of food to another, with inevitable losses at each stage. Typically cultivated fish are fed with grain or meal prepared from fish. Second, fish used to be an important source of nutrition for the poor, available locally and at little or no financial cost. Communities working together part time could harvest with simple tools the food they needed for their own use. In contrast, fish farming is undertaken to serve the markets where profits are highest. Cultivated salmon and shrimp go to the tables of the rich, not to feed the poor. And the problem is compounded by destruction of coastal fisheries. Many local stocks have been destroyed, and the prices of the rest are bid up by consumers in distant markets. As a consequence fish are becoming less available to the poor. Third, the farming of fish, shrimp, and other aquatic species causes great environmental damage. The escape of cultivated species into the wild, diffusion of food wastes and antibiotics into marine waters, the spread of viruses, and destruction of coastal wetlands are all correlates of this new technology. These harmful effects are not a random accident. They result from the working of the market because they are "externalities" that simply do not affect prices or profits in any important market for fish.

In 2002 the FAO estimated that about 75 percent of the world's oceanic fisheries were fished at or beyond capacity.[8] In 9 of the 19 world fishing zones, fish catches were above the lower limit of estimated sustainable yield.

Several high-profile events illustrate the huge stresses on world fisheries. As mentioned above, in 1992 the Canadian government closed all eastern seaboard fishing grounds, including the fisheries for cod. They remained closed in 2003, due to insufficient recovery of the stock. In 1994 salmon fishing off the U.S. West Coast was severely curtailed.[9] In the year 2002 four nations around the Caspian Sea finally agreed on a scheme to start protecting the sturgeon, the source of the famous caviar, after the annual catch had fallen from 22,000 tons a year in the 1970s to 1,000 tons a year in the late 1990s.[10] Populations of bluefin tuna, which normally live 30 years and grow to 700 kilograms (1,500 pounds), declined 94 percent in the 20 years between 1970 and 1990. The total catch from Norwegian waters is being sustained only by substituting less desirable commercial fish as the more desirable ones are being eliminated.

On the other hand, a decade-long moratorium on fisheries has rebuilt herring and cod stocks in Norwegian waters, proving that it is possible to reverse negative trends through public policy intervention. This proves harder for the European Union, which is trying to reduce the capacity of its fishing fleets. The EU fleet has increasingly shifted its presence from European waters to those of relatively poor developing countries, removing valuable jobs and protein from the local population. In sum, there is little doubt that global fisheries are pushing very hard against global limitations.

While the fishing industry around the world enjoyed fairly free and vigorous markets up to 1990, the industry experienced extraordinary techno-

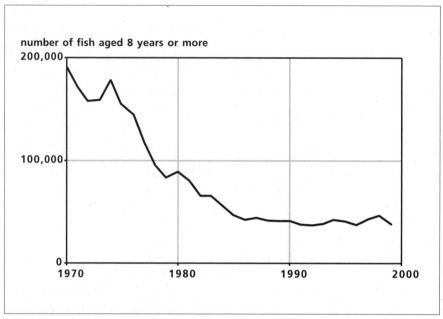

FIGURE 6-8 Bluefin Tuna Population Decline
The western Atlantic spawning population of bluefin tuna (over the age of 8) has been reduced by 80 percent over the past 30 years. Because of the high value of these fish, the fishing effort continues. (Source: ICCAT.)

logical development. Refrigerated processing boats allow fleets to stay at distant fishing grounds rather than returning home promptly with a day's catch. Radar and sonar and satellite spotting bring boats to the fish with increasing efficiency. Drift nets 30 miles long allow economical large-scale fishing even in the deep seas. The result is that harvests in more and more fisheries are overshooting the sustainable limits. Rather than protecting fish or enhancing fish stocks, the kind of technology being employed seeks to catch every last fish (figure 6-8).

Although most people understand intuitively that this leads to overexploitation of the fish stocks, the market gives no corrective feedback to keep competitors from overexploiting a common resource such as marine fish. Quite the contrary, it actively rewards those who get there first and take the most.[11] If the market signals scarcity by raising the price of fish, the richest people will be willing to pay that price. In Tokyo in the early 1990s bluefin tuna was worth as much as $100 a pound in the sushi market.[12] And in Stockholm in 2002 the price of cod, once the most ordinary sustenance of the poor, reached the unbelievable price of $80 a pound.[13] Perversely, these high prices encourage much more fishing effort among producers, as the fish population continues to be depleted. But the high price does slow the growth in demand, and it does allocate the fish to those who can pay—sadly, not those who most need it for food.

The market players who are busily exterminating resources are utterly rational. What they are doing makes complete sense, given the rewards and constraints they see from the place they occupy in the system. The fault is not with people, it is with the *system*. An unregulated market system governing a common resource with a slow regeneration rate inevitably leads to overshoot and the destruction of the commons.

> You are thinking of the whaling industry as an organization that is interested in maintaining whales; actually it is better viewed as a huge quantity of [money] capital attempting to earn the highest possible return. If it can exterminate whales in ten years and make a 15 percent profit, but it could only make 10 percent with a sustainable harvest, then it will exterminate them in ten years. After that, the money will be moved to exterminating some other resource.[14]

Only political constraints of some kind can protect the resource, and those political constraints are not easy to attain. Regulation does not necessarily work well, either. Recent research indicates that overexploitation also tends to occur when there is full private ownership of the renewable resource, and thus no opportunity for a "tragedy of the commons" syndrome.[15] Overshoot occurs simply because the information about the resource base—such as stock estimates, catch volumes, and growth rates—is uncertain and noisy and not suited to traditional management decision rules. The typical result is overinvestment in harvesting capital and overharvesting of the resource.

Traditional markets and technology have brought the globe's marine fisheries to the brink of collapse. More of the same will not restore them to health. Used with no concept of limits, markets and technologies are instruments of overshoot. Used within limits, guided by regulating institutions, however, the forces of the market and of technological development could help provide the world's fishing industry with rich harvests that can be sustained for generations.

A Summary

Exponential growth of population, capital, resource use, and pollution proceeds on the planet. It is propelled by attempts to solve keenly felt human problems, from unemployment and poverty to the need for status, power, and self-acceptance.

Exponential growth can rapidly exceed any fixed limit. If one limit is pushed back, exponential growth will soon run into another.

Because of delays in the feedback from limits, the global economic system is likely to overshoot its sustainable levels. Indeed, for many sources and sinks important to the world economy, overshoot has already occurred.

Technology and markets operate only on imperfect information and with delay. Thus they can enhance the economy's tendency to overshoot.

Technology and markets typically serve the most powerful segments of society. If the primary goal is growth, they produce growth as long as they can. If the primary goals were equity and sustainability, they could also serve those goals.

Once the population and economy have overshot the physical limits of the earth, there are only two ways back: involuntary collapse caused by escalating shortages and crises, or controlled reduction of the ecological footprint by deliberate social choice.

In the next chapter we will see what happens when technological improvements are combined with deliberate social choices to limit growth.

CHAPTER 7

Transitions to a Sustainable System

The stationary state would make fewer demands on our environmental resources, but much greater demands on our moral resources.

—HERMAN DALY, 1971

The human world can respond in three ways to signals that resource use and pollution emissions have grown beyond their sustainable limits. One way is to deny, disguise, or confuse the signals. This approach has many forms. Some claim there is no need to worry about limits; the market and technology will automatically solve any problems. Others claim there should be no attempt to reduce overshoot until there has been a great deal of additional study. Still others seek to shift the costs of their overshoot to those who are far away in space or in time. For example, it is possible to:

- Build higher smokestacks so air pollution blows farther away, where someone else has to breathe it.
- Ship toxic chemicals or nuclear waste for disposal in some distant region.
- Overharvest fish or forest resources, claiming the need to save jobs or pay debts now, while drawing down the natural stocks upon which jobs and debt payments ultimately depend.
- Subsidize extractive industries that are failing because of scarcity.
- Search for more resources while using inefficiently those already discovered.
- Compensate for falling land fertility through ever greater applications of fertilizers.
- Hold down prices by fiat or subsidies, so they cannot rise in response to scarcity.

- Use military force, or its threat, to secure the use of resources that would cost too much to purchase.

Far from solving the problems that arise from an excessive ecological footprint, these responses will only permit them to get worse.

A second way to respond is to alleviate the pressures from limits by technical or economic fixes. For example, it is possible to:

- Reduce the amount of pollution generated per mile of driving or per kilowatt of electricity generated.
- Use resources more efficiently, recycle resources, or substitute renewable resources for nonrenewable ones.
- Replace functions that nature used to perform, such as sewage treatment or flood control or soil fertilization, with energy, human capital, and labor.

These measures are urgently needed. Many of them provide increased eco-efficiency, and they will ease pressures for a while, buying essential time. But they will not eliminate the causes of those pressures. If there is less pollution per mile of driving but more driving, or more sewage treatment capacity but a rising flow of sewage, problems have only been postponed, not solved.

The third way to respond is to work on the underlying causes, to step back and acknowledge that the human socioeconomic system as currently structured is unmanageable, has overshot its limits, and is headed for collapse, and, therefore, seek to *change the structure of the system.*

The phrase *changing structure* often has ominous connotations. It has been used by revolutionaries to mean throwing people out of power, sometimes throwing bombs in the process. People may think that changing structure means changing *physical* structures, tearing down the old buildings and building new ones. Or it might be interpreted to mean changing the power structure, the hierarchy, the chain of command. Given those interpretations, changing structure appears be difficult, dangerous, and threatening to those with economic or political power.

In systems language, however, *changing structure* has little to do with throwing people out, tearing things down, or demolishing bureaucracies. In fact, doing any of those things without *real* changes in structure will just result in different people spending as much or more time and money pursuing the same goals in new buildings or organizations, producing the same old results.

In systems terms, changing structure means changing the *feedback structure, the information links* in a system: the content and timeliness of the data that actors in the system have to work with, and the ideas, goals, incentives, costs, and feedbacks that motivate or constrain behavior. The same combination of people, organizations, and physical structures can behave completely differently, if the system's actors can see a good reason for doing so, and if they have the freedom, perhaps even the incentive, to change. In time, a system with a new information structure is likely to change its social and physical structures, too. It may develop new laws, new organizations, new technologies, people with new skills, new kinds of machines or buildings. Such a transformation need not be directed centrally; it can be unplanned, natural, evolutionary, exciting, joyful.

Pervasive changes unfold spontaneously from new system structures. No one need engage in sacrifice or coercion, except, perhaps, to prevent people with vested interests from ignoring, distorting, or restricting relevant information. Human history has witnessed several structural transformations. The agricultural and industrial revolutions were the most profound examples. They both started with new *ideas* about planting food, harnessing energy, and organizing work. In fact, as we shall see in the next chapter, it is the success of those past transformations that has brought the world to the necessity for the next one, which we will call the sustainability revolution.

World3 cannot begin to represent the evolutionary dynamics of a system that is restructuring itself in new ways. But it can be used to test some of the simplest changes that might result from a society that decides to back down from overshoot and pursue goals more satisfying and sustainable than perpetual material growth.

In the previous chapter we used the World3 model to see what happens if the world makes *numerical*, not structural, changes. We put into the model higher limits, shorter delays, faster and more powerful technical responses, weaker erosion loops. If we had, instead, assumed away those structural features entirely—no limits, no delays, no erosion loops—we would have totally eliminated the overshoot and collapse behavior (as we did in Scenario 0, the "Infinity In, Infinity Out" run). But limits, delays, and erosion are physical properties of the planet. Human beings can mitigate them or enhance them, manipulate them with technologies, and accommodate them with variations in lifestyle, but humans cannot make them go away entirely.

The structural causes of overshoot over which people have the most power are the ones we did not change in chapter 6, namely those that drive the positive feedback loops causing exponential growth in human population and physical capital. They are the norms, goals, expectations, pressures, incentives, and costs that cause people to bear more than a replacement number of children. They are the deeply ingrained beliefs and practices that cause natural resources to be used more wastefully than money, that distribute income and wealth inequitably, that make people see themselves primarily as consumers and producers, that associate social status with material or financial accumulation, and that define goals in terms of getting more rather than giving more or having enough.

In this chapter we will change the positive loops that cause exponential growth in the world system. We will explore the question of how to ease down from the state of overshoot. To do that we will adopt a new perspective, focused not on technologies aimed at changing limits, but on the goals and aspirations that drive growth. We will start *only* with these positive feedback changes, without the technical changes we tested in the last chapter—and then we will put both kinds of changes together.

Deliberate Constraints on Growth

Suppose that, starting in 2002, every couple in the world understood the implications of further population growth for the welfare of their own and other children. Suppose all people were assured by their societies of acceptance, respect, material security, and care in their old age, no matter how few children they had. Suppose further that it became a shared social goal to raise every child with adequate nutrition, shelter, health care, and education. Suppose as a consequence that all couples decided to limit their family size to two surviving children (on average) and that they had readily available fertility control technologies to help them achieve that goal.

The change would entail shifts in perceived costs and benefits of children, an increased time horizon, and some concern for the welfare of others. It would give new powers, choices, and responsibilities. It would be a system restructuring equivalent to, but not the same as, the one that has already in the rich parts of the world brought down birth rates to or below replacement. It is by no means an inconceivable change; it assumes that all people adopt the reproductive choices made long ago by roughly one billion people in the most industrialized societies.

If just that change is made in World3 and no others, the results are shown in Scenario 7, figure 7-1.

To generate this scenario, we have set the average desired family size of the model population at two children and birth control effectiveness at 100 percent after the simulated year 2002. As a result, the model world's population growth slows, but age structure momentum carries the population to a peak of 7.5 billion in 2040. The peak population is half a billion lower than the maximum in Scenario 2. Thus a globally effective, two-children policy introduced in 2002 reduces the peak population less than 10 percent. The explanation is that, even without this policy, just after the turn of the millennium the model population is rapidly approaching a standard of living where small families are desired anyway and where birth control efficiencies are approaching 100 percent.

Still, reduction in the peak population has positive effects. Because of the slower population growth, consumer goods per capita, food per capita, and life expectancy are all higher than in Scenario 2. At the population peak in 2040, per capita consumer goods output is 10 percent higher, per capita food availability is 20 percent higher, and life expectancy is almost 10 percent greater than in Scenario 2. This is because less investment is needed to supply the consumption and service needs of a smaller population, so more investment is available to fuel the growth of industrial capital. As a result, industrial output grows faster and higher than it did in Scenario 2. By the year 2040 industrial output per capita has grown to twice its level in the year 2000. The model population is significantly richer than at the start of the century, and the period from 2010 to 2030 could be termed a "golden era," with relatively high human welfare for a large population.

But industrial output peaks in 2040 and declines at about the same rate as it did in Scenario 2, and for exactly the same reasons. The larger capital plant emits more pollution, which has negative effects on agricultural production. Capital has to be diverted to the agricultural sector to sustain food production. And later on, after 2050, pollution levels are sufficiently high to have negative impacts on human life expectancies. In summary, the model world experiences a "pollution crisis" where high levels of pollution poison land and lead to food shortages for the people.

Thus, given the limits and technologies assumed in the simulated world of Scenario 7, and given no constraints on material aspirations, that world cannot sustain even 7.5 billion people. We do not avoid collapse if we stabilize only the global population. Continued capital growth is as unsustainable as population growth. Each, if left unchecked, can produce an ecological footprint that exceeds the carrying capacity of the globe.

But what if the world's people decide to moderate not only their demand for children, but also their material lifestyles? What if they set themselves a goal of an adequate but not excessive standard of living? This hypothetical structural change is less visible in our present world than is the desire for fewer children, but it is certainly not unheard of.[1] It is a change advocated in nearly every religious text, a change not in the physical or political world, but in people's heads and hearts—in their goals, in their understanding of their purpose in life. To achieve this change would mean that the globe's people establish their status, derive satisfaction, and challenge themselves with goals other than ever-increasing production and ever-accumulating material wealth.

FIGURE 7-1 Scenario 7: World Seeks Stable Population from 2002
This scenario supposes that after 2002 all couples decide to limit their family size to 2 children and that they have access to effective birth control technologies. Because of age structure momentum, the population continues to grow for another generation. But the slower population growth permits industrial output to rise faster, until it is stopped by the cost of dealing with rising pollution—as in Scenario 2.

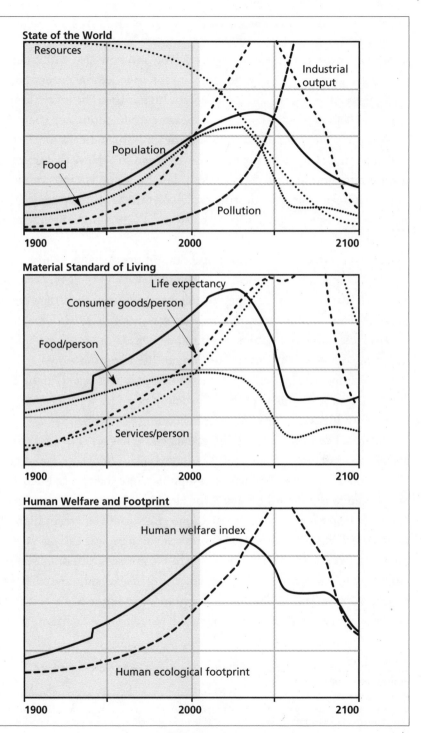

Scenario 7

Scenario 8, figure 7-2, shows a simulated world again with a desired family size of two and perfect birth control, and now also with a definition of *enough*. This world has decided to aim for an industrial output per capita level *for everyone* that is about 10 percent higher than the world average in the year 2000. In practical terms this means a tremendous step ahead for the world's poor, and a significant shift in consumption patterns for the world's rich. The model world is furthermore assumed to achieve that output with less investment, because it chooses to design capital equipment to last 25 percent longer. Average industrial capital lifetime is assumed to rise from 14 to 18 years, service capital from 20 to 25 years, agricultural inputs from 2 to 2.5 years.

As you can see from the computer output, these changes cause a considerable rise in consumer goods and services per capita in the first decade after the year 2002. In fact, they rise higher and faster than they did in the previous run, where industrial growth was not curtailed. That happens because less industrial output needs to be invested in capital growth and in replacing depreciation, given the longer lifetimes for capital. So more output is immediately available for consumption. As a result, in the decades from 2010 to 2040 this hypothetical society provides a perfectly adequate, though not luxurious, level of material comfort *for everyone.*

But this economy is not quite stabilized. It has an ecological footprint above the sustainable level, and it is forced into a long decline after 2040. The world of Scenario 8 manages to support more than seven billion people at an adequate standard of living for almost 30 years, from 2010 to 2040. Consumer goods and services per capita rise some 50 percent higher than their 2000 value. Total food production reaches a peak as early as 2010, however, and falls steadily thereafter because of stresses from pollution, which continues to rise for decades. More and more investments are made in agriculture to slow the decline in food production. For a while the capital is available, because it is not being used to achieve ever more industrial

FIGURE 7-2 Scenario 8: World Seeks Stable Population and Stable Industrial Output per Person from 2002

If the model society both adopts a desired family size of 2 children and sets a fixed goal for industrial output per capita, it can extend somewhat the "golden period" of fairly high human welfare between 2020 and 2040 in Scenario 7. But pollution increasingly stresses agricultural resources. Per capita food production declines, eventually bringing down life expectancy and population.

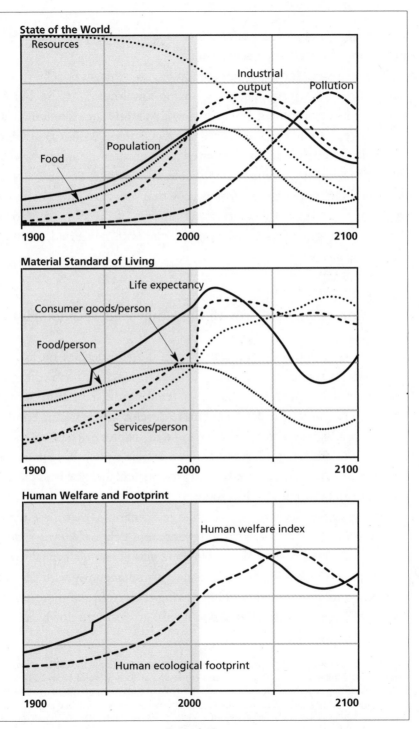

State of the World

Resources

Industrial output

Pollution

Population

Food

1900   2000   2100

Material Standard of Living

Life expectancy

Consumer goods/person

Food/person

Services/person

1900   2000   2100

Human Welfare and Footprint

Human welfare index

Human ecological footprint

1900   2000   2100

Scenario 8

growth. But gradually the burden grows to a level beyond the capacity of the industrial sector, and decline occurs.

The simulated society in this computer run manages to achieve and sustain its desired material standard of living for almost 30 years, but during that time its environment and soils steadily deteriorate. Limited consumption, limited family size, and social discipline alone do not guarantee sustainability when they are implemented too late, after the system has already overshot its limits. To remain sustainable, the world in Scenario 8 needs to do something more than control its growth. It needs to lower its ecological footprint to a level below the carrying capacity of the global environment. It needs to augment its social restructuring with concerted, appropriate, technological advance.

Constraints on Growth Plus Improved Technologies

In Scenario 9, figure 7-3, the model world again decides on an average family size of two children starting in 2002, has perfect birth control effectiveness, and sets modest limits for material production, as in Scenario 8. Furthermore, starting in 2002 it begins to develop, invest in, and employ the same technologies we tested in Scenario 6 in chapter 6. These technologies increase the efficiency of resource use, decrease pollution emissions per unit of industrial output, control land erosion, and increase land yields until food per capita reaches its desired level.

We assume in Scenario 9, as we did in Scenario 6, that these technologies become effective only after a development delay of 20 years, and that they have a capital cost. In Scenario 6 there wasn't enough capital to pay for and install the technologies while dealing simultaneously with the various crises encountered by the rapidly growing society. In the more restrained society of Scenario 9, where the population grows more slowly and capital

FIGURE 7-3 Scenario 9: World Seeks Stable Population and Stable Industrial Output per Person, and Adds Pollution, Resource, and Agricultural Technologies from 2002
In this scenario population and industrial output are limited as in the previous run, and in addition technologies are added to abate pollution, conserve resources, increase land yield, and protect agricultural land. The resulting society is sustainable: Nearly 8 billion people live with high human welfare and a continuously declining ecological footprint.

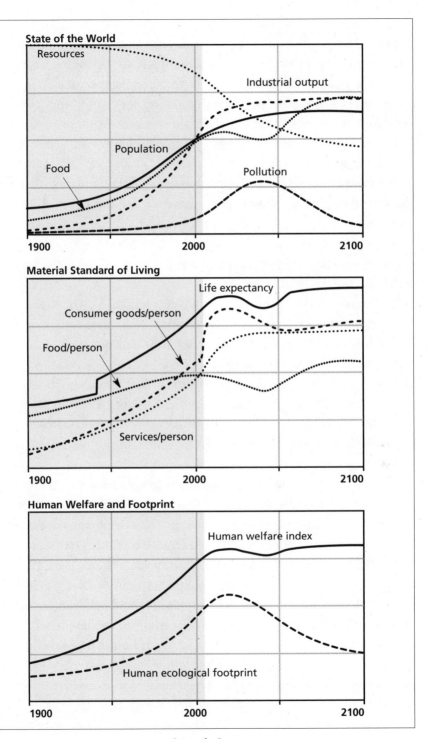

Scenario 9

does not have to fuel further growth or cope with a spiraling set of problems, the new technologies can be fully supported. Operating steadily over a century, they reduce nonrenewable resource use per unit of industrial output by 80 percent and pollution production per unit of output by 90 percent. Because growth of industrial output is contained, these gains accumulate as actual reductions in the human ecological footprint, rather than just permitting more growth.

The steady growth in land yield recedes slightly during the first half of the twenty-first century as pollution rises (a delayed effect of emissions around the end of the twentieth century—perhaps exemplified in the "real world" by the onset of global warming). But by 2040 better technologies are bringing accumulated pollution down again. Land yield recovers and rises slowly for the rest of the century.

In Scenario 9 the population levels off at less than eight billion people, who remain at their desired material standard of living throughout the century. Their life expectancy is high, though it declines slightly during the period when food production falters. Their per capita services grow to 50 percent above their level in the year 2000. By the end of the simulated twenty-first century there is sufficient food for everyone. Pollution peaks and falls before it causes irreversible damage. Nonrenewable resources deplete so slowly that nearly 50 percent of the original endowment is still present in the simulated year 2100.

The society of Scenario 9 manages to begin reducing its total burden on the environment before the year 2020; from that point the total ecological footprint of humanity is actually declining. The rate of extraction of nonrenewable resources falls after 2010. Land erosion is reduced immediately after 2002. The generation of persistent pollutants peaks a decade later. The system brings itself down below its limits, avoids an uncontrolled collapse, maintains its standard of living, and holds itself very close to equilibrium. Scenario 9 illustrates sustainability; the global system has come into equilibrium.

The word *equilibrium* in systems language means that positive and negative loops are in balance and that the system's major stocks—in this case population, capital, land, land fertility, nonrenewable resources, and pollution—are held fairly steady. It does *not* necessarily mean that the population and economy are static or stagnant. They stay roughly constant in total size, the way a river stays roughly constant in volume, even though water is always running through it. In an "equilibrium society" like the one in Scenario 9, some people are being born while others are dying; new factories, roads, buildings, machines are being built while old ones are being decommissioned and recycled. While technologies are improving, the flow of material output per person would almost certainly be changing in form, diversifying in content, increasing in quality.

As a river may rise and fall around some average flow, so could an equilibrium society vary, either by deliberate choice or by unforeseen opportunities or disasters. As a river can purify itself and support more rich and varied aquatic communities when its pollution load is diminished, so a society can purify itself of pollution, acquire new knowledge, make its production processes more efficient, shift technologies, improve its own management, make distribution more equitable, learn, and evolve. We think society is more likely to do all those things when the strains of growth are alleviated and when it is changing slowly enough that there is time for full understanding, reflection, and choice about the effects of its decisions.

The sustainable society shown in Scenario 9 is one that we believe the world could actually attain, given the knowledge about planetary systems available to us. It has nearly eight billion people, and enough food, consumer products, and services to support every one of them in comfort. It is expending considerable effort and employing continually improving technology to protect land and soils, reduce pollution, and use nonrenewable resources with high efficiency. Because its physical growth slows and eventually stops, and because its technologies work fast enough to bring its ecological footprint down to a sustainable level, it has *time, capital,* and *capacity* to solve its other problems.

We think that this is a picture not only of a feasible world, but of a desirable one. It is certainly more attractive than the simulated worlds of the previous chapter, which keep growing until they are stopped by multiple crises. Scenario 9 is not the only sustainable outcome the World3 model can produce, however. Within the system's limits there are trade-offs and choices. There could be more food and less industrial output or vice versa, more people living with a smaller ecological footprint per person, or fewer people living with more. But one principle is clear—every year of delay in starting the transition toward a sustainable equilibrium reduces the attractiveness of the trade-offs and choices that will be realistically available after the transition has been achieved. This is graphically illustrated by assuming that the policies that produced Scenario 9 were actually initiated 20 years earlier.

The Difference 20 Years Can Make

In the next run we ask: What if the model world had undertaken the sustainability policies shown in Scenario 9 (desired family size of two children, moderate material standard of living, advancing technologies of resource efficiency and pollution control) not in 2002 but in 1982? What difference does 20 years make?

Scenario 10, figure 7-4, is exactly equivalent to Scenario 9 except that the changes are made in 1982 and not in 2002. Moving to sustainability 20 years sooner could have produced a more secure and wealthy world, sooner, and with fewer adjustment problems in the agricultural sector. In this scenario the population levels off just above six billion instead of near eight billion. Pollution peaks at a much lower level and 20 years sooner, and it interferes with agriculture much less than it did in Scenario 9. Life expectancy surpasses 80 years and stays high. There are more nonrenewable resources left by the end of the twenty-first century, and it takes less effort to find and extract them. Life expectancy, food per capita, services per capita, and consumer goods per capita all end up at higher levels than they did in Scenario 9.

The Scenario 10 population is able to maintain its standard of living and support its improving technologies with no problems. This society has a more pleasant environment, more resources, more degrees of freedom; it is farther from its limits, less on the edge than the society in Scenario 9. That future might have been possible once. But the world society of 1982 did not grasp the opportunity.

FIGURE 7-4 Scenario 10: The Sustainability Policies of Scenario 9 Introduced 20 Years Earlier, in 1982

This simulation includes all the changes that were incorporated in Scenario 9, but the policies are implemented in the year 1982 instead of in 2002. Moving toward sustainability 20 years sooner would have meant a lower final population, less pollution, more nonrenewable resources, and a slightly higher average welfare for all.

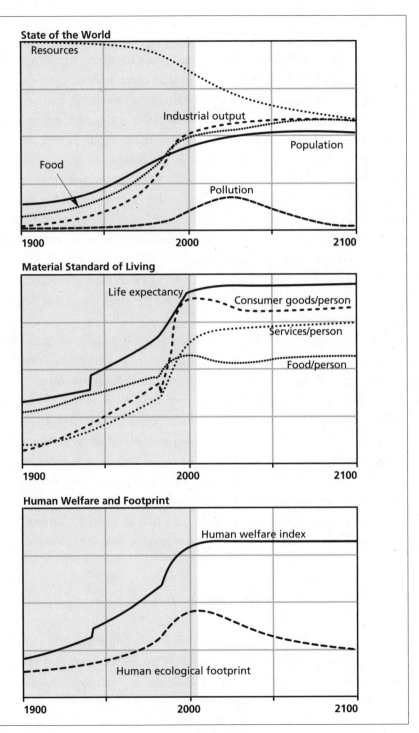

Scenario 10

We have used World3 to develop many other scenarios than the 11 reprinted here. We were exploring the possible effects of many different proposals for changes in global policies that could help ease population and the material economy back down to sustainable levels. Of course there are many simplifications and omissions in the model. So the detailed numbers produced by all these simulations are not meaningful. But there are two general insights from the effort that we do believe are valid and relevant. Our first insight from these experiments is the realization that waiting to introduce fundamental change reduces the options open for humanity's long-term future. Waiting longer to reduce population growth and stabilize productive capital stocks means that population is larger, more resources have been consumed, pollution levels are higher, more land has deteriorated, and the absolute flows of food, services, and goods required to sustain the population are higher. Needs are greater, problems larger, and capacities less.

This is nicely illustrated by implementing the policies of Scenario 9 not in 2002 but 20 years later. By then it is too late to avoid decline. Two decades of delay permit the population to rise to eight billion much sooner than it did in Scenario 9. Because of the 20-year delay in pursuing change, industrial production rises much higher than it did in Scenario 9. The added industrial activity, plus the 20-year delay in implementing pollution control technologies, brings about a pollution crisis. Pollution reduces land yield, food per capita falls, life expectancy falls, and the population declines as well. The 20-year delay in moving toward sustainability reduces the options of our simulated world, and sends it on a turbulent, and ultimately unsuccessful path. Policies that were once adequate are no longer sufficient.

How High Is Too High?

Our second insight from these experiments is perception that asking too much consumption from the global system can also produce failure. We have conducted experiments with World3 using precisely the same assumptions that produced Scenario 9, except for one change: We doubled the desired industrial output per person. The world portrayed by World3 in this case also begins to moderate its population and economy in 2002 and to implement the same resource-conserving and pollution-reducing technologies. This time, however, the model world's goal for industrial goods per capita, even with all the ameliorative technologies, cannot be sustained for the resulting population of more than seven billion people.

Industrial output per capita reaches its goal for a brief period after 2020. It peaks around 2030 and falls slowly thereafter. Food per capita declines quickly from its peak in about the same year. The reason is that too much capital is needed to attain the higher material goals and to offset the damage to the environment. By the simulated year 2050, the per capita flows of food and industrial goods available to this more ambitious world are way below what they were in the world of Scenario 9, which was content to set more moderate goals.

Does this run give us a reliable estimate for the standard of living that a "real world" of 7.5 billion people could sustain? Absolutely not! The model's numbers and its assumptions are not that reliable. No model can make precise, accurate statements about the globe 30 to 50 years in the future. It is possible that more people could actually be supported at a higher standard of living than Scenario 9. It is also possible, given the optimistic assumptions in World3 about no war, no conflict, no corruption, and no mistakes, that the consumption level shown in Scenario 9 could never actually be supported sustainably.

World3 serves, in some ways, like an architect's sketch. It shows the interrelationship among important variables. It helps us think, in general terms, about the future in which we wish to live. But it does not provide any details about the complex political, psychological, and personal issues involved in constructing the transition. Planning those requires expertise beyond ours. And in the event, it would require experimentation, humility, openness to information about mistakes, and the willingness to adjust course during the process.

We do not necessarily infer from our experiments with the model that implementation now of sustainable policies will lead to an attractive future while a delay of 10 or 20 years will ordain society to failure. But we do conclude that delays are reducing the levels of affluence we could eventually enjoy sustainably. We do not infer from our scenarios that a consumption goal equal to today's or 10 percent or 20 percent above current levels is sustainable, while a consumption goal twice that is a recipe for disaster. But we do conclude that a sustainable system could offer consumption standards attractive to many in the world today. On the other hand it can not sustainably offer unlimited, or even very high, levels of material consumption to a population of 6 to 8 billion.

World3 cannot be used to fine-tune a human world seeking to find and live at its upper sustainable limits. No model now available, probably no model ever available, will permit that kind of numerical precision. Furthermore, maximization of the human ecological footprint is a dangerous policy, since the actual physical limits to growth are variable and uncertain, and we will always learn of them and respond to them only after delays. It would be safer, and probably preferable for other reasons as well, to learn to live satisfying lives safely below the globe's estimated limits, rather than always straining to achieve the maximum that is physically possible.

World3 is a model designed to explore the behavior modes of an interconnected, nonlinear, delayed-response, limited system. It is not intended to spell out an exact prediction for the future or a detailed plan for action. But the runs shown in this chapter suggest general conclusions that we believe are valid and that are not at all recognized in the public discourse. Imagine how differently decisions would be made, investments would be allocated, news would be reported, laws would be debated, if the following information were widely known and accepted:

- A global transition to a sustainable society is probably possible without reductions in either population or industrial output.
- A transition to sustainability will require, however, an active decision to reduce the human ecological footprint. This, in turn, will require personal decisions to reduce family size, lower goals for industrial growth, and raise efficiency in use of the earth's resources.
- There are many ways in which a sustainable society could be

structured, many choices about numbers of people, living standards, technological investments, and allocations among industrial goods, services, food, and other material needs. These
choices need not be made the same way in every part of the
world, but they do need to be made soon.

- There are unavoidable trade-offs between the number of people
  the earth can sustain and the material level at which each person
  can be supported. The exact numerical trade-offs are not know-
  able. They will change over time as technology, knowledge,
  human coping ability, and the earth's support systems change.
  Even so, the general implication remains: More people will mean
  less sustainable material throughput and a smaller ecological
  footprint for each person.

- The longer the world economy takes to reduce its ecological
  footprint and move toward sustainability, the lower the population and material standard that will be ultimately supportable. At
  some point, delay means collapse.

- The higher the society sets its targets for population and material
  standard of living, the greater are its risks of exceeding and
  eroding its limits.

According to our computer model, our mental models, our knowledge
of the data, and our experience of the "real world," there is no time to waste
in easing down below the limits and setting goals for sustainability. Putting
off the reduction of throughputs and the transition to sustainability means,
at best, diminishing the options of future generations, and, at worst, precip-
itating a collapse.

There is no reason to waste time, either. Sustainability is a new idea to
many people, and many find it hard to understand. But all over the world
there are people who have entered into the exercise of imagining and
bringing into being a sustainable world. They see it as a world to move
toward not reluctantly, but joyfully, not with a sense of sacrifice, but a sense
of adventure. A sustainable world could be very much better than the one
we live in today.

The Sustainable Society

There are many ways to define *sustainability*. It is simplest to say that a sustainable society is one that can persist over generations; one that is farseeing enough, flexible enough, and wise enough not to undermine either its physical or its social systems of support.

In 1987 the World Commission on Environment and Development put the idea of sustainability into memorable words:

> A sustainable society is one that "meets the needs of the present without compromising the ability of future generations to meet their own needs."[2]

From a systems point of view, a sustainable society is one that has in place informational, social, and institutional mechanisms to keep in check the positive feedback loops that cause exponential population and capital growth. This means that birth rates roughly equal death rates, and investment rates roughly equal depreciation rates, unless or until technical changes and social decisions justify a considered, limited change in the levels of population or capital. In order to be socially sustainable, the combination of population and capital and technology would have to be configured so that the material living standard is adequate and secure for everyone and fairly distributed. To be materially and energetically sustainable, the economy's throughputs would have to meet Herman Daly's three conditions:[3]

- Its rates of use of renewable resources do not exceed their rates of regeneration.
- Its rates of use of nonrenewable resources do not exceed the rate at which sustainable renewable substitutes are developed.
- Its rates of pollution emission do not exceed the assimilative capacity of the environment.

Such a society, with a sustainable ecological footprint, would be almost unimaginably different from the one in which most people now live. Mental models at the start of the twenty-first century are imprinted by powerful images of persistent poverty or of rapid material growth and determined efforts to maintain that growth at all costs. Dominated by images of heedless growth or frustrating stagnation, the shared human consciousness can hardly envision a purposeful, sufficient, just, and sustainable society. Before we can elaborate here on what sustainability *could* be, we need to start with what it *need not* be.

Sustainability need not mean "zero growth." A society fixated on growth tends to shun any questioning of that goal, but questioning growth does not have to mean denying growth. As Aurelio Peccei, founder of the Club of Rome, pointed out in 1977, that would just substitute one oversimplification for another:

> All those who had helped to shatter the myth of growth . . . were ridiculed and figuratively hanged, drawn, and quartered by the loyal defenders of the sacred cow of growth. Some of those . . . accuse the [*Limits to Growth*] report . . . of advocating ZERO GROWTH. Clearly, such people have not understood anything, either about the Club of Rome, or about growth. The notion of zero growth is so primitive—as, for that matter, is that of infinite growth—and so imprecise, that it is conceptual nonsense to talk of it in a living, dynamic society.[4]

A sustainable society would be interested in qualitative development, not physical expansion. It would use material growth as a considered tool, not a perpetual mandate. Neither for nor against growth, it would begin to discriminate among kinds of growth and purposes for growth. It could even entertain rationally the idea of purposeful negative growth, to undo excess, to get below limits, to cease doing things that, in a full accounting of natural and social costs, actually cost more than they are worth.

Before a sustainable society would decide on any specific growth proposal, it would ask what the growth is for, and who would benefit, and what it would cost, and how long it would last, and whether the growth could be accommodated by the sources and sinks of the earth. Such a society would apply its values and its best knowledge of the earth's limits to choose only those kinds of growth that would serve important social goals while enhancing sustainability. Once any physical growth had accomplished its purposes, society would stop its pursuit.

A sustainable society would *not* paralyze into permanence the current inequitable patterns of distribution. It would certainly not lock the poor permanently in their poverty. To do so would not be sustainable for two reasons. First, the poor would not and should not stand for it. Second, keeping any part of the population in poverty would not, except under dire coercive measures or rising death rates, stabilize the population. For both practical and moral reasons, any sustainable society must provide sufficiency and security for all. To get to sustainability from here, the remaining growth possible—whatever space there is for more resource use and pollution emissions, plus whatever space is freed up by higher efficiencies and lifestyle moderations on the part of the rich—would logically and, one would hope, joyfully be allocated to those who need it most.

A sustainable state would *not* be a society of despondency and stagnancy, unemployment and bankruptcy that current economic systems experience when their growth is interrupted. The difference between a sustainable society and a present-day economic recession is like the difference between stopping an automobile purposely with the brakes versus stopping it by crashing into a brick wall. When the present economy overshoots, it turns around too quickly and unexpectedly for people or enterprises to retrain, relocate, or readjust. A deliberate transition to sustainability would take place slowly enough, and with enough forewarning, so that people and businesses could find their places in the new economy.

There is no reason why a sustainable society need be technically or culturally primitive. Freed from both anxiety and greed, it would have enormous possibilities for human creativity. Without the high costs of growth for both society and environment, technology and culture could bloom. John Stuart Mill, one of the first (and last) economists to take seriously the idea of an economy consistent with the limits of the earth, saw that what he called a "stationary state" could support an evolving and improving society. More than 150 years ago he wrote:

> I cannot . . . regard the stationary state of capital and wealth with the unaffected aversion so generally manifested towards it by polit-

ical economists of the old school. I am inclined to believe that it would be, on the whole, a very considerable improvement on our present condition. I confess I am not charmed with the ideal of life held out by those who think that the normal state of human beings is that of struggling to get on; that the trampling, crushing, elbowing, and treading on each other's heels . . . are the most desirable lot of humankind. . . . It is scarcely necessary to remark that a stationary condition of capital and population implies no stationary state of human improvement. There would be as much scope as ever for all kinds of mental culture and moral and social progress; as much room for improving the Art of Living, and much more likelihood of its being improved.[5]

A sustainable world would not and could not be a rigid one, with population or production or anything else held pathologically constant. One of the strangest assumptions of present-day mental models is the idea that a world of moderation must be a world of strict, centralized government control. For a sustainable economy, that kind of control is not possible, desirable, or necessary. (From a systems point of view, it has serious deficiencies, as the former Soviet Union amply demonstrated.)

A sustainable world would need rules, laws, standards, boundaries, social agreements, and social constraints, of course, as does every human culture. Some of the rules for sustainability would be different from the rules people are used to now. Some of the necessary controls are already coming into being, as, for example, in the international ozone agreement and the greenhouse gas negotiations. But rules for sustainability, like every workable social rule, would be put into place not to destroy freedoms, but to create freedoms or to protect them. A ban on bank robbing inhibits the freedom of the thief in order to assure that everyone else has the freedom to deposit and withdraw money safely. A ban on overuse of a renewable resource or on the generation of a dangerous pollutant protects vital freedoms in a similar way.

It doesn't take much imagination to come up with a minimum set of social structures—feedback loops that carry new information about costs, consequences, and sanctions—that would allow evolution, creativity, and change, and permit many more freedoms than would ever be possible in a world that continues to crowd against or exceed its limits. One of the most important of these new rules would fit in perfectly with economic theory: It would combine knowledge and regulation to "internalize the externalities" of the market system, so that the price of a product would reflect the full costs (including all environmental and social side effects) of making that product. This is a measure every economics textbook has called for (in vain) for decades. It would automatically guide investments and purchases, so people could make choices in the monetary realm that they would not later regret in the realm of real material or social worth.

Some people think that a sustainable society would have to stop using nonrenewable resources, since their use is by definition unsustainable. That idea is an over-rigid interpretation of what it means to be sustainable. Certainly a sustainable society would use nonrenewable gifts from the earth's crust more thoughtfully and efficiently than the present world does. It would price them properly, thereby keeping more of them available for future generations. But there is no reason not to use them, as long as their use meets the criteria of sustainability already defined, namely that they do not overwhelm a natural sink and that renewable substitutes be developed.

There is no reason for a sustainable society to be uniform. As in nature, diversity in a human society would be both a cause of and a result of sustainability. Some people who have thought about sustainability envision it as largely decentralized, with localities relying more on their local resources and less on international trade. They would set boundary conditions that keep each community from threatening the viability of the others or of the Earth as a whole. Cultural variety, autonomy, freedom, and self-determination could be greater, not less, in such a world.

There is no reason for a sustainable society to be undemocratic or boring or unchallenging. Some games that amuse and consume people today, such as arms races or the accumulation of unlimited wealth, would probably no longer be feasible, respected, or interesting. But there still would be games, challenges, problems to solve, ways for people to prove themselves, serve each other, test their abilities, and live good lives—perhaps more satisfying lives than any possible today.

That was a long list of what a sustainable society is *not*. In the process of spelling it out, we have also, by contrast, implied what we think a sustainable society could be. But the details of that society will not be worked out by one small bunch of computer modelers; this will require the ideas, visions, and talents of billions of people.

From the structural analysis of the world system we have described in this book, we can contribute only a simple set of general guidelines for restructuring any system toward sustainability. We list them below. Each one can be worked out in hundreds of ways at all levels—households, communities, corporations, nations, and the world as a whole. Some people will see how to implement these guidelines in their own lives and cultures and political and economic systems. One step in any of these directions is a step toward sustainability, though ultimately all the steps must be taken.

- *Extend the planning horizon.* Base the choice among current options much more on their long-term costs and benefits, not just the results they will produce in today's market or tomorrow's election. Develop the incentives, the tools, and the procedures required for the media, the market, and elections to report, respect, and be responsible for issues that unfold over decades.
- *Improve the signals.* Learn more about and monitor both the real welfare of the human population and the real impact on the world ecosystem of human activity.[6] Inform governments and the public as continuously and promptly about environmental and social conditions as about economic conditions. Include environmental and social costs in economic prices; recast economic indicators such as the GDP, so that they do not confuse costs with benefits or throughput with welfare or the deterioration of natural capital with income.
- *Speed up response times.* Look actively for signals that indicate when the environment or society is stressed. Decide in advance what to do if problems appear (if possible, forecast them before they appear) and have in place the institutional and technical arrangements necessary to act effectively. Educate for flexibility and creativity, for critical thinking and the ability to redesign both physical and social systems. Computer modeling can help with

this step, but equally important would be general education in systems thinking.

- *Minimize the use of nonrenewable resources.* Fossil fuels, fossil groundwaters, and minerals should be used only with the greatest possible efficiency, recycled when possible (fuels can't be recycled, but minerals and water can), and consumed only as part of a deliberate transition to renewable resources.

- *Prevent the erosion of renewable resources.* The productivity of soils, surface waters, rechargeable groundwaters, and all living things, including forests, fish, and game should be protected and, as far as possible, restored and enhanced. These resources should only be harvested at the rate at which they can regenerate themselves. That requires information about their regeneration rates and strong social sanctions or economic inducements against their overuse.

- *Use all resources with maximum efficiency.* The more human welfare can be obtained within a given ecological footprint, the better the quality of life can be while remaining below the limits. Great efficiency gains are both technically possible and economically favorable.[7] Higher efficiency will be essential, if the current world population and economy are to get back down below the limits without inducing a collapse.

- *Slow and eventually stop exponential growth of population and physical capital.* There are limits to the extent that the first six items on this list can be pursued. Therefore this last item is the most essential. It involves institutional and philosophical change and social innovation. It requires defining levels of population and industrial output that are desirable and sustainable. It calls for defining goals around the idea of development rather than growth. It asks, simply but profoundly, for a larger and more truly satisfying vision of the purpose of human existence than mere physical expansion and accumulation.

We can expand on this last, important step toward sustainability by acknowledging the pressing problems that underlie much of the cultural commitment to growth: poverty, unemployment, and unmet needs. Growth as presently structured either is not solving these problems at all, or is solving them only slowly and inefficiently. Until more effective solutions are in sight, however, society will never let go of its addiction to growth, because people so badly need hope. Growth may be a false hope, but it is better than no hope at all.

To restore hope and to solve very real problems, these are three areas where completely new thinking is needed.

- *Poverty.* Sharing is a forbidden word in political discourse, probably because of the deep fear that real sharing would mean not enough for anyone. "Sufficiency" and "solidarity" are concepts that can help structure new approaches to ending poverty. We are all in this overshoot together. There is enough to go around, if we manage well. If we don't manage well, no one, no matter how wealthy, will escape the consequences.
- *Unemployment.* Human beings need to work, to test and to discipline themselves, to take responsibility for fulfilling their own basic needs, to have the satisfaction of personal participation, and to be accepted as adult, responsible members of society. That need should be not be left unfulfilled, and it should not be filled by degrading or harmful work. At the same time, employment should not be a requirement for the ability to subsist. Creativity is needed here to get beyond the narrow idea that some people "create" jobs for others, and the even narrower idea that workers are simply costs to be cut. What is needed is an economic system that uses and supports the contributions all people are able to make, that shares work, leisure, and economic outputs equitably, and that does not abandon people who for reasons temporary or permanent cannot work.
- *Unmet nonmaterial needs.* People don't need enormous cars; they need admiration and respect. They don't need a constant stream of new clothes; they need to feel that others consider them to be attractive, and they need excitement and variety and beauty. People don't need electronic entertainment; they need something

interesting to occupy their minds and emotions. And so forth. Trying to fill real but nonmaterial needs—for identity, community, self-esteem, challenge, love, joy—with material things is to set up an unquenchable appetite for false solutions to never-satisfied longings. A society that allows itself to admit and articulate its nonmaterial human needs, and to find nonmaterial ways to satisfy them, would require much lower material and energy throughputs and would provide much higher levels of human fulfillment.

How, in practice, can anyone attack these problems? How can the world evolve a *system* that solves them? That is the opportunity for creativity and choice. The generations that live around the turn of the twenty-first century are called upon not only to bring their ecological footprint below the earth's limits, but to do so while restructuring their inner and outer worlds. That process will touch every arena of life, require every kind of human talent. It will need technical and entrepreneurial innovation, as well as communal, social, political, artistic, and spiritual invention. Fifty years ago Lewis Mumford recognized the magnitude of the task and its uniquely human character; it is one that will challenge and develop the *humanity* of everyone.

An age of expansion is giving place to an age of equilibrium. The achievement of this equilibrium is the task of the next few centuries. . . . The theme for the new period will be neither arms and the man nor machines and the man: its theme will be the resurgence of life, the displacement of the mechanical by the organic, and the re-establishment of the person as the ultimate term of all human effort. Cultivation, humanization, co-operation, symbiosis: these are the watchwords of the new world-enveloping culture. Every department of life will record this change: it will affect the task of education and the procedures of science no less than the organization of industrial enterprises, the planning of cities, the development of regions, the interchange of world resources.[8]

The necessity of taking the industrial world to its next stage of evolution is not a disaster—it is an amazing opportunity. How to seize the opportunity, how to bring into being a world that is not only sustainable, functional, and equitable but also deeply *desirable* is a question of leadership and ethics and vision and courage, properties not of computer models but of the human heart and soul. To speak of them we—the authors—need a chapter break here. We need to turn off our computers, put away our data and scenarios, and reappear in chapter 8, where we will conclude with insights that have come as much from our hearts and our intuition as they have from our scientific analyses.

CHAPTER 8

Tools for the Transition to Sustainability

We must be careful not to succumb to despair, for there is still the
odd glimmer of hope.

—EDOUARD SAOUMA, 1993

Can we move nations and people in the direction of sustainability?
Such a move would be a modification of society comparable in scale
to only two other changes: the Agricultural Revolution of the late
Neolithic and the Industrial Revolution of the past two centuries.
Those revolutions were gradual, spontaneous, and largely uncon-
scious. This one will have to be a fully conscious operation, guided
by the best foresight that science can provide. . . . If we actually do
it, the undertaking will be absolutely unique in humanity's stay on
the Earth.

—WILLIAM D. RUCKELSHAUS, 1989

We have been writing about, talking about, and working toward sustainability for over three decades now. We have had the privilege of
knowing thousands of colleagues in every part of the world who work in
their own ways, with their own talents, in their own societies toward a sustainable society. When we act at the official, institutional level and when we
listen to political leaders, we often feel frustrated. When we work with individuals, we usually feel encouraged.

Everywhere we find folks who care about the earth, about other people,
and about the welfare of their children and grandchildren. They recognize
the human misery and the environmental degradation around them, and
they question whether policies that promote more growth along the same
old lines can make things better. Many of them have a feeling, often hard for

them to articulate, that the world is headed in the wrong direction and that preventing disaster will require some big changes. They are willing to work for those changes, if only they could believe their efforts would make a positive difference. They ask: *What can I do? What can governments do? What can corporations do? What can schools, religions, media do? What can citizens, producers, consumers, parents do?*

Experiments guided by those questions are more important than any specific answers, though answers abound. There are "50 simple things you can do to save the planet." Buy an energy-efficient car, for one. Recycle your bottles and cans, vote knowledgeably in elections—if you are among those people in the world blessed with cars, bottles, cans, or elections. There are also not-so-simple things to do: Work out your own frugally elegant lifestyle, have at most two children, argue for higher prices on fossil energy (to encourage energy efficiency and stimulate development of renewable energy), work with love and partnership to help one family lift itself out of poverty, find your own "right livelihood," care well for one piece of land, do whatever you can to oppose systems that oppress people or abuse the earth, run for election yourself.

All these actions will help. And, of course, they are not enough. Sustainability and sufficiency and equity require structural change; they require a revolution, not in the political sense, like the French Revolution, but in the much more profound sense of the agricultural or industrial revolutions. Recycling is important, but by itself it will not bring about a revolution.

What will? In search of an answer, we have found it helpful to try to understand the first two great revolutions in human culture, insofar as historians can reconstruct them.

The First Two Revolutions: Agriculture and Industry

About 10,000 years ago the human population, after millennia of evolution, had reached the huge (for the time) number of about 10 million. These people lived as nomadic hunter-gatherers, but in some regions their numbers had begun to overwhelm the once abundant plants and game. To adapt to the problem of disappearing wild resources they did two things. Some of them intensified their migratory lifestyle. They moved out of their ancestral homes in Africa and the Middle East and populated other areas of the game-rich world.

Others started domesticating animals, cultivating plants, and *staying in one place*. That was a totally new idea. Simply by staying put, the proto-farmers altered the face of the planet, the thoughts of humankind, and the shape of society in ways they could never have foreseen.

For the first time it made sense to own land. People who didn't have to carry all their possessions on their backs could accumulate things, and some could accumulate more than others. The ideas of wealth, status, inheritance, trade, money, and power were born. Some people could live on excess food produced by others. They could become full-time toolmakers, musicians, scribes, priests, soldiers, athletes, or kings. Thus arose, for better or worse, guilds, orchestras, libraries, temples, armies, competitive games, dynasties, and cities.

As its inheritors, we think of the agricultural revolution as a great step forward. At the time it was probably a mixed blessing. Many anthropologists think that agriculture was not a better way of life, but a necessary one to accommodate increasing populations. Settled farmers got more food from a hectare than hunter-gatherers did, but the food was of lower nutritional quality and less variety, and it required much more work to produce. Farmers became vulnerable in ways nomads never were to weather, disease, pests, invasion by outsiders, and oppression from their emerging ruling classes. People who did not move away from their own wastes experienced humankind's first chronic pollution.

Nevertheless, agriculture was a successful response to wildlife scarcity. It permitted yet more population growth, which added up over centuries to an enormous increase, from 10 million to 800 million people by 1750. The larger population created new scarcities, especially in land and energy. Another revolution was necessary.

The industrial revolution began in England with the substitution of abundant coal for vanishing trees. The use of coal raised practical problems of earthmoving, mine construction, water pumping, transport, and controlled combustion. These problems were solved relatively quickly, resulting in concentrations of labor around mines and mills. The process elevated technology and commerce to a prominent position in human society—above religion and ethics.

Again everything changed in ways that no one could have imagined. Machines, not land, became the central means of production. Feudalism gave way to capitalism and to capitalism's dissenting offshoot, communism. Roads, railroads, factories, and smokestacks appeared on the landscape. Cities swelled. Again the change was a mixed blessing. Factory labor was even harder and more demeaning than farm labor. The air and waters near the new factories turned unspeakably filthy. The standard of living for most of the industrial workforce was far below that of a farmer. But farmland was not available; work in a factory was.

It is hard for people alive today to appreciate how profoundly the industrial revolution changed human thought, because that thought still shapes our perceptions. In 1988 historian Donald Worster described the philosophical impact of industrialism perhaps as well as any of its inheritors and practitioners can:

> The capitalists . . . promised that, through the technological domination of the earth, they could deliver a more fair, rational, efficient and productive life for everyone. . . . Their method was simply to free individual enterprise from the bonds of traditional hierarchy and community, whether the bondage derived from other humans or the earth . . . That meant teaching everyone to treat the earth, as well as each other, with a frank, energetic, self-assertiveness. . . . People must . . . think constantly in terms of making money. They must regard everything around them—the land, its natural resources, their own labor—as potential commodities that might fetch a profit in the market. They must demand the right to produce, buy, and sell those commodities without outside regulation or interference. . . . As wants multiplied, as markets grew more and more far-flung, the bond between humans and the rest of nature was reduced to the barest instrumentalism.[1]

That bare instrumentalism led to incredible productivity and a world that now supports, at varying levels of sufficiency, 6,000 million people—

more than 600 times the population existing before the agricultural revolution. Far-flung markets and swelling demands drive environmental exploitation from the poles to the tropics, from the mountaintops to the ocean depths. The success of the industrial revolution, like the previous successes of hunting-gathering and of agriculture, eventually created its own scarcity, not only of game, not only of land, not only of fuels and metals, but of the total carrying capacity of the global environment. Humankind's ecological footprint had once more exceeded what was sustainable. Success created the necessity for another revolution.

The Next Revolution: Sustainability

It is as impossible now for anyone to describe the world that could evolve from a sustainability revolution as it would have been for the farmers of 6000 BC to foresee the corn and soybean fields of modern Iowa, or for an English coal miner of AD 1800 to imagine an automated Toyota assembly line. Like the other great revolutions, the coming sustainability revolution will also change the face of the land and the foundations of human identities, institutions, and cultures. Like the previous revolutions, it will take centuries to unfold fully—though it is already under way.

Of course no one knows how to bring about such a revolution. There is not a checklist: "To accomplish a global paradigm shift, follow these 20 steps." Like the great revolutions that came before, this one can't be planned or dictated. It won't follow a list of fiats from government or a proclamation from computer modelers. The sustainability revolution will be organic. It will arise from the visions, insights, experiments, and actions of billions of people. The burden of making it happen is not on the shoulders of any one person or group. No one will get the credit, but everyone can contribute.

Our systems training and our own work in the world have affirmed for us two properties of complex systems germane to the sort of profound revolution we are discussing here.

First, information is the key to transformation. That does not necessarily mean *more* information, better statistics, bigger databases, or the World Wide Web, though all of these may play a part. It means *relevant, compelling, select, powerful, timely, accurate* information flowing in new ways to new recipients, carrying new content, suggesting new rules and goals (rules and goals that are themselves information). When its information flows are changed, any system will behave differently. The policy of *glasnost*, for example—the simple opening of information channels that had long been closed in the Soviet Union—guaranteed the rapid transformation of Eastern Europe beyond anyone's expectations. The old system had been held in place by tight control of information. Letting go of that control triggered total system restructuring (turbulent and unpredictable, but inevitable).

Second, systems strongly resist changes in their information flows, especially in their rules and goals. It is not surprising that those who benefit from the current system actively oppose such revision. Entrenched political, economic, and religious cliques can constrain almost entirely the attempts of an individual or small group to operate by different rules or to attain goals different from those sanctioned by the system. Innovators can be ignored, marginalized, ridiculed, denied promotions or resources or public voices. They can be literally or figuratively snuffed out.

Only innovators, however—by perceiving the need for new information, rules, and goals, communicating about them, and trying them out—can make the changes that transform systems. This important point is expressed clearly in a quote that is widely attributed to Margaret Mead, "Never deny the power of a small group of committed individuals to change the world. Indeed that is the only thing that ever has."

We have learned the hard way that it is difficult to live a life of material moderation within a system that expects, exhorts, and rewards consumption. But one can move a long way in the direction of moderation. It is not easy to use energy efficiently in an economy that produces energy-inefficient products. But one can search out, or if necessary invent, more efficient ways of doing things, and in the process make those ways more accessible to others.

Above all, it is difficult to put forth new information in a system that is structured to hear only old information. Just try, sometime, to question in public the value of more growth, or even to make a distinction between growth and development, and you will see what we mean. It takes courage and clarity to challenge an established system. But it can be done.

In our own search for ways to encourage the peaceful restructuring of a system that naturally resists its own transformation, we have tried many tools. The obvious ones are displayed through this book—rational analysis, data gathering, systems thinking, computer modeling, and the clearest words we can find. Those are tools that anyone trained in science and economics would automatically grasp. Like recycling, they are useful, necessary, and they are not enough.

We don't know what will be enough. But we would like to conclude by mentioning five other tools we have found *helpful*. We introduced and discussed this list for the first time in our 1992 book. Our experience since then has affirmed that these five tools are not optional; they are essential characteristics for any society that hopes to survive over the long term. We present them here again in our concluding chapter "not as *the* ways to work toward sustainability, but as *some* ways."

"We are a bit hesitant to discuss them," we said in 1992, " because we are not experts in their use and because they require the use of words that do not come easily from the mouths or word processors of scientists. They are considered too 'unscientific' to be taken seriously in the cynical public arena."

What are the tools we approached so cautiously?

They are: visioning, networking, truth-telling, learning, and loving.

It seems like a feeble list, given the enormity of the changes required. But each of these exists within a web of positive loops. Thus their persistent and consistent application initially by a relatively small group of people would have the potential to produce enormous change—even to challenge the present system, perhaps helping to produce a revolution.

"The transition to a sustainable society might be helped," we said in 1992, "by the simple use of words like these more often, with sincerity and without apology, in the information streams of the world." But we used them with apology ourselves, knowing how most people would receive them.

Many of us feel uneasy about relying on such "soft" tools when the future of our civilization is at stake, particularly since we do not know how to summon them up, in ourselves or in others. So we dismiss them and turn the conversation to recycling or emission trading or wildlife preserves or some other necessary but insufficient part of the sustainability revolution—but at least a part we know how to handle.

So let's talk about the tools we don't yet know how to use, because humanity must quickly master them.

Visioning

Visioning means imagining, at first generally and then with increasing specificity, what you really want. That is, *what you really want*, not what someone has taught you to want, and not what you have learned to be willing to settle for. Visioning means taking off the constraints of "feasibility," of disbelief and past disappointments, and letting your mind dwell upon its most noble, uplifting, treasured dreams.

Some people, especially young people, engage in visioning with enthusiasm and ease. Some find the exercise of visioning frightening or painful, because a glowing picture of what *could be* makes what *is* all the more intolerable. Some people never admit their visions, for fear of being thought impractical or "unrealistic." They would find this paragraph uncomfortable to read, if they were willing to read it at all. And some people have been so crushed by their experience that they can only explain why any vision is impossible. That's fine; skeptics are needed, too. Vision needs to be disciplined by skepticism.

We should say immediately, for the sake of the skeptics, that we do not believe vision makes anything happen. Vision without action is useless. But action without vision is directionless and feeble. Vision is absolutely necessary to guide and motivate. More than that, vision, when widely shared and firmly kept in sight, does *bring into being new systems*.

We mean that literally. Within the limits of space, time, materials, and energy, visionary human intentions can bring forth not only new information, new feedback loops, new behavior, new knowledge, and new technology, but also new institutions, new physical structures, and new powers within human beings. Ralph Waldo Emerson recognized this profound truth 150 years ago:

> Every nation and every man instantly surround themselves with a
> material apparatus which exactly corresponds to their moral state,

or their state of thought. Observe how every truth and every error, each a thought of some man's mind, clothes itself with societies, houses, cities, language, ceremonies, newspapers. Observe the ideas of the present day . . . see how each of these abstractions has embodied itself in an imposing apparatus in the community, and how timber, brick, lime, and stone have flown into convenient shape, obedient to the master idea reigning in the minds of many persons. . . .

It follows, of course, that the least change in the man will change his circumstances; the least enlargement of ideas, the least mitigation of his feelings in respect to other men . . . would cause the most striking changes of external things.[2]

A sustainable world can never be fully realised until it is widely envisioned. The vision must be built up by many people before it is complete and compelling. As a way of encouraging others to join in the process, we'll list here some of what we see when we let ourselves imagine a sustainable society we would like to live in—as opposed to one we would be willing to settle for. This is by no means a definitive list. We include it here only to invite you to develop and enlarge it.

- Sustainability, efficiency, sufficiency, equity, beauty, and community as the highest social values.
- Material sufficiency and security for all. Therefore, by individual choice as well as communal norms, low birth rates and stable populations.
- Work that dignifies people instead of demeaning them. Some way of providing incentives for people to give their best to society and to be rewarded for doing so, while ensuring that everyone will be provided for sufficiently under any circumstances.
- Leaders who are honest, respectful, intelligent, humble, and more interested in doing their jobs than in keeping their jobs, more interested in serving society than in winning elections.

- An economy that is a means, not an end, one that serves the welfare of the environment, rather than vice versa.
- Efficient, renewable energy systems.
- Efficient, closed-loop materials systems.
- Technical design that reduces emissions and waste to a minimum, and social agreement not to produce emissions or waste that technology and nature can't handle.
- Regenerative agriculture that builds soils, uses natural mechanisms to restore nutrients and control pests, and produces abundant, uncontaminated food.
- The preservation of ecosystems in their variety, with human cultures living in harmony with those ecosystems; therefore, high diversity of both nature and culture, and human appreciation for that diversity.
- Flexibility, innovation (social as well as technical), and intellectual challenge. A flourishing of science, a continuous enlargement of human knowledge.
- Greater understanding of whole systems as an essential part of each person's education.
- Decentralization of economic power, political influence, and scientific expertise.
- Political structures that permit a balance between short-term and long-term considerations; some way of exerting political pressure now on behalf of our grandchildren.
- High-level skills on the part of citizens and governments in the arts of nonviolent conflict resolution.
- Media that reflect the world's diversity and at the same time unite cultures with relevant, accurate, timely, unbiased, and intelligent information, presented in its historic and whole-system context.
- Reasons for living and for thinking well of ourselves that do not involve the accumulation of material things.

Networking

We could not do our work without networks. Most of the networks we belong to are informal. They have small budgets, if any, and few of them appear on rosters of world organizations.[3] They are almost invisible, but their effects are not negligible. Informal networks carry information in the same way as formal institutions do, and often more effectively. They are the natural home of new information, and out of them new system structures can evolve.[4]

Some of our networks are very local, some are international. Some are electronic, some involve people looking each other in the face every day. Whatever their form, they are made up of people who share a common interest in some aspect of life, who stay in touch and pass around data and tools and ideas and encouragement, who like and respect and support each other. One of the most important purposes of a network is simply to remind its members that they are not alone.

A network is nonhierarchical. It is a web of connections among equals, held together not by force, obligation, material incentive, or social contract, but by shared values and the understanding that some tasks can be accomplished together that could never be accomplished separately.

We know of networks of farmers who share organic pest control methods. There are networks of environmental journalists, "green" architects, computer modelers, game designers, land trusts, consumer cooperatives. There are thousands and thousands of networks that developed as people with common purposes found each other. Some networks become so busy and essential that they evolve into formal organizations with offices and budgets, but most come and go as needed. The advent of the World Wide Web certainly has facilitated and accelerated the formation and maintenance of networks.

Networks dedicated to sustainability at both the local and the global levels are especially needed to create a sustainable society that harmonizes with local ecosystems while keeping itself within global limits. About local networks we can say little here; our localities are different from yours. One role of local networks is to help reestablish the sense of community and relation to place that has been largely lost since the industrial revolution.

When it comes to global networks, we would like to make a plea that they be truly global. The means of participation in international information streams are as badly distributed as are the means of production. There are more telephones in Tokyo, it has been said, than in all of Africa. That must be even more true of computers, fax machines, airline connections, and invitations to international meetings. But once more the wonder of human inventiveness seems to provide a surprising solution in the form of the Web and cheap access devices.

One could argue that Africa and other underrepresented parts of the world should attend first to their needs for many things other than computers and Web access. We disagree; the needs of the underprivileged cannot be effectively communicated, nor can the world benefit from their contributions, unless their voices can be heard. Some of the greatest gains in material and energy efficiency have come in the design of communications equipment. It is possible within a sustainable ecological footprint for everyone to have the opportunity for global as well as local networking. We must close the "Digital Divide."

If some part of the sustainability revolution interests you, you can find or form a network of others who share your particular interests. The network will help you discover where to go for information, what publications and tools are available, where to find administrative and financial support, and who can help with specific tasks. The right network will not only help you learn, but also allow you to pass your learning on to others.

Truth-Telling

We are no more certain of the truth than anyone is. But we often know an untruth when we hear one. Many untruths are deliberate, understood as such by both speaker and listeners. They are put forth to manipulate, lull, or entice, to postpone action, to justify self-serving action, to gain or preserve power, or to deny an uncomfortable reality.

Lies distort the information stream. A system cannot function well if its information streams are corrupted by lies. One of the most important tenets of systems theory, for reasons we hope we have made clear in this book, is that information should not be distorted, delayed, or sequestered.

"All of humanity is in peril," said Buckminster Fuller, "if each one of us does not dare, now and henceforth, always to tell only the truth and all the truth, and to do so promptly—right now."[5] Whenever you speak to anyone, on the street, at work, to a crowd, and especially to a child, you can endeavor to counter a lie or affirm a truth. You can deny the idea that having more things makes one a better person. You can question the notion that more for the rich will help the poor. The more you can counter misinformation, the more manageable our society will become.

Here are some common biases and simplifications, verbal traps, and popular untruths that we run into frequently in discussing limits to growth. We think they need to be pointed out and avoided, if there is ever to be clear thinking about the human economy and its relationship to a finite Earth.

Not: A warning about the future is a prediction of doom.
But: A warning about the future is a recommendation to follow a different path.

Not: The environment is a luxury or a competing demand or a commodity that people will buy when they can afford it.
But: The environment is the source of all life and every economy. Opinion polls typically show that the public is willing to pay more for a healthy environment.

Not: Change is sacrifice, and it should be avoided.
But: Change is challenge, and it is necessary.

Not: Stopping growth will lock the poor in their poverty.
But: It is the avarice and indifference of the rich that lock the poor into poverty. The poor need new attitudes among the rich; then there will be growth specifically geared to serve their needs.

Not: Everyone should be brought up to the material level of the richest countries.

But: There is no possibility of raising material consumption levels for everyone to the levels now enjoyed by the rich. Everyone should have their fundamental material needs satisfied. Material needs beyond this level should be satisfied only if it is possible, for all, within a sustainable ecological footprint.

Not: All growth is good, without question, discrimination, or investigation.
Nor: All growth is bad.
But: What is needed is not growth, but development. Insofar as development requires physical expansion, it should be equitable, affordable, and sustainable, with all real costs counted.

Not: Technology will solve all problems.
Nor: Technology does nothing but cause problems.
But: We need to encourage technologies that will reduce the ecological footprint, increase efficiency, enhance resources, improve signals, and end material deprivation.
And: We must approach our problems as human beings and bring more to bear on them than just technology.

Not: The market system will automatically bring us the future we want.
But: We must decide for ourselves what future we want. Then we can use the market system, along with many other organizational devices, to achieve it.

Not: Industry is the cause of all problems, or the cure.
Nor: Government is the cause or the cure.
Nor: Environmentalists are the cause or the cure.
Nor: Any other group [economists come to mind] is the cause or the cure.
But: All people and institutions play their role within the large system structure. In a system that is structured for overshoot, all players deliberately or inadvertently contribute to that overshoot. In a system that is structured for sustainability, industries, governments, environmentalists, and most especially economists will play essential roles in contributing to sustainability.

Not: Unrelieved pessimism.

Nor: Sappy optimism.

But: The resolve to tell the truth about both the successes and failures of the present and the potentials and obstacles in the future.

And above all: The courage to admit and bear the pain of the present, while keeping a steady eye on a vision of a better future.

Not: The World3 model, or any other model, is right or wrong.

But: All models, including the ones in our heads, are a little right, much too simple, and mostly wrong. How do we proceed in such a way as to test our models and learn where they are right and wrong? How do we speak to each other as fellow modelers with an appropriate mixture of skepticism and respect? How do we stop playing right–wrong games with each other and start designing right–wrong tests for our models against the real world?

That last challenge, sorting out and testing models, brings us to the topic of learning.

Learning

Visioning, networking, and truth-telling are useless if they do not inform action. There are many things to *do* to bring about a sustainable world. New farming methods have to be worked out. New businesses have to be started and old ones have to be redesigned to reduce their footprint. Land has to be restored, parks protected, energy systems transformed, international agreements reached. Laws have to be passed and others repealed. Children have to be taught, and so do adults. Films have to be made, music played, books published, Web sites established, people counseled, groups led, subsidies removed, sustainability indicators developed, and prices corrected to portray full costs.

All people will find their own best role in all this doing. We wouldn't presume to prescribe a specific role for anyone but ourselves. But we would make one suggestion: Whatever you do, do it humbly. Do it not as immutable policy, but as experiment. Use your action, whatever it is, to learn.

The depths of human ignorance are much more profound than most of us are willing to admit. This is especially so at a time when the global economy is coming together as a more integrated whole than it has ever been, when that economy is pressing against the limits of a wondrously complex planet, and when wholly new ways of thinking are called for. At this time, no one knows enough. No leaders, no matter how authoritative they pretend to be, understand the situation. No policy should be imposed wholesale upon the whole world. If you cannot afford to lose, do not gamble.

Learning means the willingness to go slowly, to try things out, and to collect information about the effects of actions, including the crucial but not always welcome information that the action is not working. One can't learn without making mistakes, telling the truth about them, and moving on. Learning means exploring a new path with vigor and courage, being open to other people's explorations of other paths, and being willing to switch paths if one is found that leads more directly to the goal.

The world's leaders have lost both the habit of learning and the freedom to learn. Somehow a political system has evolved in which the voters expect leaders to have all the answers, that assigns only a few people to be leaders, and that brings them down quickly if they suggest unpleasant remedies. This perverse system undermines the leadership capacity of the people and the learning capacity of the leaders.

It's time for us to do some truth-telling on this issue. The world's leaders do not know any better than anyone else how to bring about a sustainable society; most of them don't even know it's necessary to do so. A sustainability revolution requires each person to act as a learning leader at some level, from family to community to nation to world. And it requires each of us to support leaders by allowing them to admit uncertainty, conduct honest experiments, and acknowledge mistakes.

No one can be free to learn without patience and forgiveness. But in a condition of overshoot, there is not much time for patience and forgiveness. Finding the right balance between the apparent opposites of urgency and patience, accountability and forgiveness is a task that requires compassion, humility, clearheadedness, honesty, and—that hardest of words, that seemingly scarcest of all resources—love.

Loving

One is not allowed in the industrial culture to speak about love, except in the most romantic and trivial sense of the word. Anyone who calls upon the capacity of people to practice brotherly and sisterly love, love of humanity as a whole, love of nature and of our nurturing planet, is more likely to be ridiculed than to be taken seriously. The deepest difference between optimists and pessimists is their position in the debate about whether human beings are able to operate collectively from a basis of love. In a society that systematically develops individualism, competitiveness, and short-term focus, the pessimists are in the vast majority.

Individualism and shortsightedness are the greatest problems of the current social system, we think, and the deepest cause of unsustainability. Love and compassion institutionalized in collective solutions is the better alternative. A culture that does not believe in, discuss, and develop these better human qualities suffers from a tragic limitation in its options. "How good a society does human nature permit?" asked psychologist Abraham Maslow. "How good a human nature does society permit?"[6]

The sustainability revolution will have to be, above all, a collective transformation that permits the best of human nature, rather than the worst, to be expressed and nurtured. Many people have recognized that necessity and that opportunity. For example, John Maynard Keynes wrote in 1932:

> The problem of want and poverty and the economic struggle between classes and nations is nothing but a frightful muddle, a transitory and unnecessary muddle. For the Western World already has the resource and the technique, if we could create the organization to use them, capable of reducing the Economic Problem, which now absorbs our moral and material energy, to a position of secondary importance. . . .

Thus the . . . day is not far off when the Economic Problem will take the back seat where it belongs, and . . . the arena of the heart and head will be occupied . . . by our real problems—the problems of life and of human relations, of creation and behaviour and religion.[7]

Aurelio Peccei, the great industrial leader who wrote constantly about problems of growth and limits, economics and environment, resources and governance, never failed to conclude that the answers to the world's problems begin with a "new humanism." In 1981 he expressed this view:

The humanism consonant with our epoch must replace and reverse principles and norms that we have heretofore regarded as untouchable, but that have become inapplicable, or discordant with our purpose; it must encourage the rise of new value systems to redress our inner balance, and of new spiritual, ethical, philosophical, social, political, aesthetic, and artistic motivations to fill the emptiness of our life; it must be capable of restoring within us . . . love, friendship, understanding, solidarity, a spirit of sacrifice, conviviality; and it must make us understand that the more closely these qualities link us to other forms of life and to our brothers and sisters everywhere in the world, the more we shall gain.[8]

It is not easy to practice love, friendship, generosity, understanding, or solidarity within a system whose rules, goals, and information streams are geared for lesser human qualities. But we try, and we urge you to try. Be patient with yourself and others as you and they confront the difficulty of a changing world. Understand and empathize with inevitable resistance; there is resistance, some clinging to the ways of unsustainability, within each of us. Seek out and trust in the best human instincts in yourself and in everyone. Listen to the cynicism around you and have compassion for those who believe in it, but don't believe it yourself.

Humanity cannot triumph in the adventure of reducing the human footprint to a sustainable level if that adventure is not undertaken in a spirit of global partnership. Collapse cannot be avoided if people do not learn to view themselves and others as part of one integrated global society. Both will require compassion, not only with the here and now, but with the distant and future as well. Humanity must learn to love the idea of leaving future generations a living planet.

Is anything we have advocated in this book, from more resource efficiency to more compassion, really possible? Can the world actually ease down below the limits and avoid collapse? Can the human footprint be reduced in time? Is there enough vision, technology, freedom, community, responsibility, foresight, money, discipline, and love, on a global scale?

Of all the hypothetical questions we have posed in this book, these are the most unanswerable, though many people will pretend to answer them. Even we—your authors—differ among ourselves when tallying the odds for and against. The ritual cheerfulness of many uninformed people, especially world leaders, would say the questions are not even relevant; there are no meaningful limits. Many of the informed are infected with the deep cynicism that lies just under the ritual public cheerfulness. They would say that there are severe problems already, with worse ones ahead, and that there's not a chance of solving them.

Both of those answers are based, of course, on mental models. The truth of the matter is that *no one knows*.

We have said many times in this book that the world faces not a preordained future, but a choice. The choice is between different mental models, which lead logically to different scenarios. One mental model says that this world for all practical purposes has no limits. Choosing that mental model will encourage extractive business as usual and take the human economy even farther beyond the limits. The result will be collapse.

Another mental model says that the limits are real and close, and that there is not enough time, and that people cannot be moderate or responsible or compassionate. At least not in time. That model is self-fulfilling. If the world's people choose to believe it, they will be proven right. The result will be collapse.

A third mental model says that the limits are real and close and in some cases below our current levels of throughput. But there is just enough time, with no time to waste. There is just enough energy, enough material, enough money, enough environmental resilience, and enough human virtue to bring about a planned reduction in the ecological footprint of humankind: a sustainability revolution to a much better world for the vast majority.

That third scenario might very well be wrong. But the evidence we have seen, from world data to global computer models, suggests that it could conceivably be made right. There is no way of knowing for sure, other than to try it.

Changes from World3 to World3-03

To prepare the scenarios presented in this book, we used an updated version of the computer model World3-91.

World3 was originally constructed for use in our 1972 volume, the first edition of *The Limits to Growth*. It was fully described in the technical report of our study.[1] The model was originally written in a computer simulation language called DYNAMO. By 1990, a new language, STELLA, offered the best tools for our analysis. When we prepared the scenarios for our 1992 volume *Beyond the Limits*, the World3 model was converted from DYNAMO to STELLA and updated into a new version called World3-91. The changes we made for that conversion are described in the appendix to *Beyond the Limits*.[2]

When preparing the scenarios for the current book, it proved useful to update World3-91 slightly. The resulting model, called World3-03, is available on a CD-ROM.[3] But it is simple to summarize the few changes necessary to convert World3-91 into World3-03. Three of the changes calculate the cost of technology in a different way; one change makes desired family size respond more strongly to growth in industrial output. The other changes have no influence on the behavior of the model; they just make it easier to understand its behavior. The changes are:

- Change the determinant of the capital cost of new technology in three sectors. The capital cost should be determined by *implemented* technology, not *available* technology—in the resources, pollution, and agricultural sectors.
- Change a lookup table in the population sector to make desired family size slightly more responsive to high levels of industrial output per capita.
- Add the new variable called *human welfare index*—an indicator of the well-being of the average global citizen. The definition of this index is provided in appendix 2.

- Add the new variable called *human ecological footprint*—an indicator of humanity's total environmental burden on the planet. The definition of this index is provided in appendix 2.
- Change the plot scale for population—to simplify reading.
- Define a new graph that shows the behavior of the human welfare index and the human ecological footprint for the period between 1900 and 2100.

To assist the reader we provide here STELLA flow diagrams for the new structures. We also describe the plot scales used for the scenarios in this book. The full STELLA equation list of World 3-03 and other information are provided on the CD-ROM.

New Structures in World3-03

The STELLA flow chart for the new technology formulation is shown below, exemplified with land yield technology. This formulation is replicated for the resource and the pollution sectors.

industrial capital output ratio multiplier from land yield technology

technology development delay

land yield technology   land yield technology change rate

land yield multiplier from technology

land yield technology change rate multiplier

food ratio

desired food ratio

When the model variable food ratio (food per person/subsistence food per person) falls below desired levels, World3 begins to develop technologies that raise the yield on the land. Analogous formulations provide enhanced technology when the resources required per unit of industrial output rise above desired levels, and when the pollution generated per unit of output rises above desired levels.

The STELLA flow chart for HWI, human welfare index, is shown below. The underlying logic is described in appendix 2.

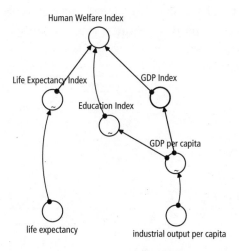

The STELLA flowchart for the HEF, human ecological footprint, is shown below. The underlying logic is described in appendix 2.

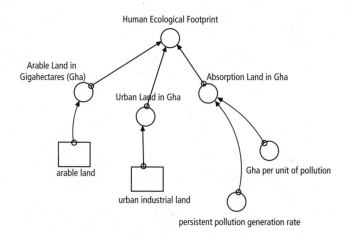

The World3-03 Scenario Scales

The values of 11 variables from World3-03 are plotted in the three graphs presented for each scenario in this book. We did not put numerical scales on the vertical axes of these graphs, because we do not consider the precise values of the variables in each scenario to be very significant. We do provide those scales here, however, for readers with a more technical interest in the simulations. The 11 variables are plotted on very different scales, but those scales are held constant throughout the 11 scenarios:

Graph 1: State of the World

Variable	Low value	High value
Population	0	12×10^9
Total food production	0	6×10^{12}
Total industrial production	0	4×10^{12}
Index of persistent pollution	0	40
Nonrenewable resources	0	2×10^{12}

Graph 2: Material Standard of Living

Variable	Low value	High value
Food per capita	0	1,000
Consumer goods per capita	0	250
Services per capita	0	1,000
Life expectancy	0	90

Graph 3: Human Welfare and Ecological Footprint

Variable	Low value	High value
Human welfare indicator	0	1
Human ecological footprint	0	4

Indicators of Human Welfare and Ecological Footprint

Background

When discussing the future of humanity on planet Earth, it is helpful to define two concepts—"human welfare" and the "human ecological footprint." These describe respectively the quality of life of the average global citizen in its broadest sense, including both material and immaterial components, and the total environmental impact placed on the global resource base and ecosystem by humanity.

Both concepts are simple to grasp in principle, but hard to define with precision. And the limitations on available time series data force us to make significant approximations when we develop mathematical equations to express them. But, generally speaking, human welfare increases when anyone increases personal satisfaction without anyone else reducing theirs. The human ecological footprint increases when there is an increase in resource extraction, pollution emission, land erosion, or biodiversity destruction, without a simultaneous reduction in other human impacts on nature.

To illustrate the use of the two concepts, let us paraphrase the ideal we have pursued in this book as follows: It is to increase "human welfare" while ensuring that the "ecological footprint" is as small as possible, and—at the very least—stays below what the global ecosystem can sustain in the very long run—the global carrying capacity.

Many analysts have spent much time and effort trying to create operational indicators for both human welfare and the ecological footprint. GDP per capita is often used as a simple measure of welfare, even though it has gross shortcomings in that role. World2,[1] the historical predecessor of World3, included a hotly debated "Quality of Life Index" that took into consideration the effect on human welfare of four factors: crowding, food, pollution, and material consumption.

After considering the options, we chose the indicators described below. We chose quantitative indicators, since they best suit the mathematical model World3. And instead of defining our own indices, we chose to adapt existing indicators that are more generally accepted.

The Human Development Index of UNDP

As a measure of human welfare we chose the Human Development Index (HDI), which has been measured for a number of years for most countries by the United Nations Development Program (UNDP). The HDI is published annually in the *Human Development Report*.[2] In the 2001 *Report* UNDP defined the HDI:

> The HDI is a summary measure of human development. It measures the average achievement in a country in three basic dimensions of human development:
>
> - A long and healthy life, as measured by life expectancy at birth
> - Knowledge, as measured by the adult literacy rate (with two-thirds weight) and the combined primary, secondary and tertiary gross enrolment rate (with one-third weight)
> - A decent standard of living, as measured by GDP per capita (in PPP-$, Purchasing Power Parity US dollars)[3]

The UNDP calculates the HDI as the arithmetic average of three indices (the life expectancy index, the education index, and the GDP index)—one for each of the three factors listed in the quote above.

The life expectancy and education indices increase linearly with life expectancy, literacy, and school enrollment. The GDP index also increases when the GDP per capita increases. But in the latter case, the UNDP assumes strongly diminishing returns once GDP per capita exceeds the 1999 level of former Eastern European countries.[4]

The Human Welfare Index in World3

As a measure of human welfare in World3 we formulated a variable we call the human welfare index (HWI). The HWI approximates the UNDP's HDI—to the extent this is possible using only the variables in the World3 model. The resulting STELLA flow diagram is shown in appendix 1, and the detailed formulation is available on the World3-03 CD ROM.

The human welfare index in World3 is the sum of the life expectancy, education, and GDP indices, divided by three. The resulting HWI grows from around 0.2 in the year 1900 to 0.7 in the year 2000. It reaches a maximum of 0.8 in the most successful scenarios around 2050. These three values are equivalent to the HDI in 1999 of respectively Sierra Leone, Iran, and the Baltic republics.

The value of our HWI in year 1999 is very close to the actual HDI calculated by the UNDP for that year, which stood at 0.71 for the world average.[5]

The Ecological Footprint of Mathis Wackernagel

As a measure of the "human ecological footprint" we adapted the ecological footprint (EF) as developed by Mathis Wackernagel and colleagues in the 1990s. Wackernagel et al. calculated the ecological footprint for a number of countries,[6] and in some cases time series showing the change over time of the footprint of individual countries. Highly relevant for the purposes of this work, Wackernagel also calculated the ecological footprint of the global population and its development from 1961 to 1999.[7] The ecological footprint of most nations of the world is published biannually by World Wide Fund for Nature.[8]

Wackernagel defines his ecological footprint as the land area necessary to provide for the current way of life. His EF is measured in (global average) hectares. He adds up the amount of cropland, grazing land, forestland, fishing grounds, and built-up land needed to maintain a given population (country, region, world) at a given lifestyle. He adds the amount of forest land that would have been required to absorb the carbon dioxide emitted from the fossil energy used by the population. All types of land are then converted to land of average biological productivity. The number of "average hectares" is calculated by the use of a scaling factor, which is proportional to the biological productivity of the land (the ability of the land to produce biomass). Wackernagel wants to extend his framework to include the land necessary to neutralize emissions (other gases, toxics) and land needed for freshwater use, but he has not yet been able to do so in a meaningful manner.

The biological productivity of a piece of land depends on what technology is used. Intense use of fertilizer will ensure a larger crop from the same hectare. Thus more fertilizer will reduce the EF—unless the CO_2 emissions from the production of the fertilizer require more (absorption) land than what was saved through the increased yield. Since technologies change continuously, Wackernagel's land productivities do the same—in pace with the "average technology" in use at the time.[9]

Thus the EF increases when humanity uses larger areas for food or fiber, or emits more CO_2. Even if the latter emissions are not absorbed in forests (and instead accumulate in the atmosphere), the footprint—the area that would have been necessary for absorption purposes if CO_2 were not to accumulate in the atmosphere—grows. This is how overshoot is possible, until the accumulation of greenhouse gases forces a change in human behavior that reduces the EF.

The Human Ecological Footprint in World3

As a measure of the human ecological footprint in the World3 model, we have formulated the index we call the human ecological footprint (HEF). The HEF approximates Wackernagel's ecological footprint to the extent that this is possible within the confines of the limited number of variables in the World3 model. The resulting STELLA flow diagram is shown in appendix 1, and the detailed formulation is available on the World3-03 CD ROM.

The human ecological footprint in World3 is the sum of three components: the arable land used for crop production in agriculture; the urban land used for urban–industrial–transportation infrastructure; and the amount of absorption land required to neutralize the emission of pollutants, assumed to be proportional to the persistent pollution generation rate. All land areas are measured in billion (10^9) hectares.

The HEF is normalized to be 1 in 1970, and the resulting index varies from 0.5 in the year 1900, through 1.76 in year 2000, to highly unsustainable values above 3 for short periods in scenarios showing overshoot and collapse. In the most successful scenarios it proves possible to keep the HEF below 2 for most of the twenty-first century. The sustainable level of HEF is probably around 1.1, a level that was passed around 1980.

Endnotes

Authors' Preface

1. Donella H. Meadows, Dennis L. Meadows, Jorgen Randers, and William W. Behrens III, *The Limits to Growth* (New York: Universe Books, 1972).
   There were also 2 technical books: Dennis L. Meadows et al., *The Dynamics of Growth in a Finite World* (Cambridge, MA: Wright-Allen Press, 1974), and Dennis L. Meadows and Donella H. Meadows, *Toward Global Equilibrium* (Cambridge, MA: Wright-Allen Press, 1973). The first is a complete documentation of the World3 computer model; the second presents 13 chapters with auxiliary studies and submodels made as input to the global model. Both books are now distributed by Pegasus Communications, One Moody Street, Waltham, MA 02453-5339 (www.pegasuscom.com).

2. Donella H. Meadows, Dennis L. Meadows, and Jorgen Randers, *Beyond the Limits* (Post Mills, VT: Chelsea Green Publishing Company, 1992).

3. Yes, there was a World1 and also a World2. World1 was the prototype model first sketched out by MIT Professor Jay Forrester in response to the Club of Rome's inquiry about interconnections among global trends and problems. World2 is Forrester's final documented model, described in Jay W. Forrester, *World Dynamics* (Cambridge, MA: Wright-Allen Press, 1971). This book is now distributed by Pegasus Communications. The World3 model was developed from World2, primarily by elaborating its structure and extending its quantitative database. Forrester is the intellectual father of the World3 model and of the system dynamics modelling method it employs.

4. See the *Report of the World Summit on Sustainable Development*, United Nations, A/CONF.199/20, New York, 2002 (also available on www.un.org), which includes the targets agreed to in the Plan of Implementation; for example, to halve by 2015 the number of people who lack access to clean water and sanitation, to reduce the global loss of biodiversity by 2010, and to restore global fisheries to maximum sustainable yield by 2015. In spite of the level of concern reflected in these commitments, in the eyes of some NGO observers the WSSD did not make much progress, and in some cases even backtracked on commitments made in Rio 10 years earlier.

5. World Commission on Environment and Development, *Our Common Future* (Oxford: Oxford University Press, 1987). The commission is widely known as the Brundtland Commission, after its leader, Gro Harlem Brundtland, former Prime Minister of Norway. In LTG we used the word "equilibrium" instead of "sustainability."

6. The World Bank, *World Bank Atlas–2003*, Washington, DC, 2003, 64–65.

7. Mathis Wackernagel et al., "Tracking the ecological overshoot of the human economy,"

Proceedings of the Academy of Science, 99, no. 14:9266–9271, Washington, DC, 2002. Also available at www.pnas.org/cgi/doi/10.1073/pnas.142033699.

8. See Meadows et al., *The Dynamics of Growth in a Finite World,* 501 and 57, for the LTG numbers, which match the actual numbers in Lester Brown et al., *Vital Signs 2000* (New York: W. W. Norton, 2000), 99.

9. See Meadows et al., *The Dynamics of Growth in a Finite World,* 501 and 264, for the LTG numbers, which show an increase of 67% from 1972 to 2000, matching well the 63% increase in world grain production reported in Brown, *Vital Signs 2000, 35.*

Chapter 1. Overshoot

1. M. Wackernagel et al., "Ecological Footprints of Nations: How Much Nature Do They Use? How Much Nature Do They Have?" (Xalapa, Mexico: Centro de Estudios para la Sustentabilidad [Center for Sustainability Studies], March 10, 1997). See also Mathis Wackernagel et al., "Tracking the Ecological Overshoot of the Human Economy," *Proceedings of the Academy of Science* 99, no. 14 (Washington, DC, 2002): 9266–9271. Also available at www.pnas.org/cgi/doi/10.1073/pnas.142033699.

2. World Wide Fund for Nature, *Living Planet Report 2002* (Gland, Switzerland: WWF, 2002).

3. The comparison includes *all* scenarios except two (Scenarios 0 and 10), which illustrate purely hypothetical worlds.

4. U Thant, 1969.

5. "World Scientists' Warning to Humanity," December 1992, available from Union of Concerned Scientists, 26 Church Street, Cambridge, MA 02238. Also available at www.ucsusa.org/ucs/about/page.cfm?pageID=1009.

6. "Making Sustainable Commitments: An Environmental Strategy for the World Bank" (discussion draft) (Washington, DC: World Bank, April 17, 2001), xii.

7. World Commission on Environment and Development, *Our Common Future* (Oxford: Oxford University Press, 1987), 8.

Chapter 2. The Driving Force: Exponential Growth

1. This exercise is described by Linda Booth-Sweeney and Dennis Meadows, *The Systems Thinking Playbook,* vol. 3 (Durham, NH: University of New Hampshire, 2001), 36–48.

2. We are indebted to Robert Lattes for telling us this riddle.

3. This approximation gives useful values for doubling times only when there is frequent compounding of the interest. For example, a growth rate of 100 percent per day would give a doubling time of about 0.72 day—17 hours—if the growing quantity were incremented by 4.17 percent per hour. But if the increment occurs only once a day, as in the peanut example cited below, the doubling time is one day.

4. World Bank, *The Little Data Book 2001* (Washington, DC: World Bank, 2001), 164.

5. Population Reference Bureau, *1998 World Population Data Sheet.*

6. United Nations Population Division, *1998 Revision: World Population Estimates and Projections* (New York: United Nations Department of Economic and Social Affairs, 1998).

7. PRB, *1998 Data Sheet.*

8. The gross national income (GNI) equals the gross domestic product (GDP) plus the nation's income from abroad. GDP is the money value of the production of goods and services inside the national border.

9. See, for example, Partha S. Dasgupta, "Population, Poverty and the Local Environment," *Scientific American,* February 1995, 40; Bryant Robery, Shea O. Rutstein, and Leo Morris, "The Fertility Decline in Developing Countries," *Scientific American,* December 1993, 60; and Griffith Feeney, "Fertility Decline in East Asia," *Science* 266 (December 2, 1994), 1518.

10. For the details, see Donella H. Meadows, "Population Sector," in D. L. Meadows et al., *Dynamics of Growth in a Finite World* (Cambridge, MA: Wright-Allen Press, 1974).

11. This confusion was illustrated by a story told to us sometime in the early 1970s by the great geologist M. King Hubbert. During the Second World War the British, knowing that the Japanese were about to invade the Malay peninsula—source of the world's rubber—undertook a massive effort to move all the rubber they could find to a safe stockpile in India. They just managed, as the Japanese moved in, to accumulate enough rubber in India to supply them, they hoped, with tires and other essential rubber products for the duration of the war. But one night the rubber stockpile caught fire and was totally destroyed.

    "That's okay," said some British economists upon receiving the news.

    "It was insured."

12. See William W. Behrens III, Dennis L. Meadows, and Peter M. Milling, "Capital Sector," in *Dynamics of Growth in a Finite World.*

13. John C. Ryan and Alan Thein Durning, *Stuff: The Secret Lives of Everyday Things* (Seattle: Northwest Environment Watch, 1997), 46.

14. World Bank, *World Development Indicators—2001* (Washington, DC: World Bank, 2001), 4.

15. United Nations Development Program, *Human Development Report 1998* (New York and Oxford: Oxford University Press, 1998), 29.

16. Ibid., 2.

17. See, for example, Peter Senge, *The Fifth Discipline* (New York: Doubleday, 1990), 385–386.

18. We implicitly model "success to the successful" loops by assuming the world's current distribution patterns, unless we intervene to change them.

19. Lester R. Brown, Gary Gardner, and Brian Halweil, "Beyond Malthus: Sixteen Dimensions of the Population Problem," *Worldwatch Paper 143* (Washington, DC: Worldwatch Institute, September 1998).

Chapter 3. The Limits: Sources and Sinks

1. Herman Daly, "Toward Some Operational Principles of Sustainable Development," *Ecological Economics* 2 (1990): 1–6. See a further elaboration in the introduction to Herman Daly, *Beyond Growth* (Boston: Beacon Press, 1996).

2. For a recent, thorough, and systematic review of the most imminent global limits, see Lester Brown, *Eco-Economy* (New York: W. W. Norton, 2001), chapters 2 and 3. For a broad review of—and data on—global physical limits, see World Resources Institute, *World Resources 2000–2001: People and Ecosystems: The Fraying Web of Life* (Oxford: Elsevier Science Ltd., 2002), part 2, "Data Tables."

3. Even more ways of aiding and accelerating the transition to sustainability are presented in a systematic manner in Brown, *Eco-Economy*, chapters 4–12.

4. Lester R. Brown, "Feeding Nine Billion," from Lester R. Brown et al., *State of the World 1999* (New York: W. W. Norton, 1999), 118.

5. Calculated by us, assuming a subsistence need of 230 kilograms (506 pounds) of grain per person per year.

6. WRI, *World Resources 1998–99*, 155.

7. United Nations Food and Agriculture Organization, *The Sixth World Food Survey* (Rome: FAO, 1996).

8. P. Pinstrup-Anderson, R. Pandya-Lorch, and M. W. Rosengrant, *1997, The World Food Situation: Recent Developments, Emerging Issues, and Long-Term Prospects* (Washington, DC: International Food Policy Research Institute, 1997).

9. Lester R. Brown, Michael Renner, and Brian Halweil, *Vital Signs 1999* (New York: W. W. Norton, 1999), 146.

10. G. M. Higgins et al., *Potential Population Supporting Capacities of Lands in the Developing World* (Rome: FAO, 1982). This technical study is summarized in a nontechnical report by Paul Harrison, *Land, Food, and People* (Rome: FAO, 1984). This factor of 16 is based on extremely optimistic assumptions, and it applies only to developing countries, which are starting from a low yield base. The FAO has not done a similar study for the lands of the industrialized countries.

11. Sara J. Scherr, "Soil Degradation: A Threat to Developing-Country Food Security by 2020?" *IFPRI Discussion Paper 27* (Washington, DC: IFPRI, February 1999), 45.

12. Food from the sea is even more limited than land-based food, and its use is even more obviously beyond sustainable limits. Futuristic schemes for non-land-based food—aquaculture, yeast in vats, and so forth—will be marginal as major food sources, mainly because of the energy and capital they demand and the pollution they produce. Food not grown primarily on land using photosynthesis from the sun's energy would be even more unsustainable than the present food system. Genetically modified crops, at least so far, seem to be developed for pest resistance or herbicide resistance in order to cut down on expensive inputs, rather than to increase yields.

13. For an excellent summary of global soil loss studies, see Scherr, "Soil Degradation."

14. United Nations Environment Program, "Farming Systems Principles for Improved Food Production and the Control of Soil Degradation in the Arid, Semi-Arid, and Humid Tropics," proceedings of an expert meeting cosponsored by the International Crops Research Institute for the Semi-Arid Tropics, Hyderabad, India, 1986.

15. B. G. Rosanov, V. Targulian, and D. S. Orlov, "Soils," in *The Earth as Transformed by Human Action: Global and Regional Changes in the Biosphere Over the Past 30 Years*, edited by B. L. Turner et al. (Cambridge: Cambridge University Press, 1990). See also Brown, *Eco-Economy*, 62–68.

16. L. R. Oldeman, "The Global Extent of Soil Degradation," in *Soil Resilience and Sustainable*

Land Use, edited by D. J. Greenland and T. Szaboles (Wallingford, UK: Commonwealth Agricultural Bureau International, 1994).

17. All figures in this paragraph are from Gary Gardner, "Shrinking Fields: Cropland Loss in a World of Eight Billion," *Worldwatch Paper 131* (Washington, DC: Worldwatch Institute, 1996).

18. WRI, *World Resources 1998–99*, 157. It is estimated that soil degradation between 1945 and 1990 has reduced global food output by 17 percent of what it would otherwise have been.

19. Quotes by Cassman, Ruttan, and Loomis are from Charles C. Mann, "Crop Scientists Seek a New Revolution," *Science* 283 (January 15, 1999): 310.

20. For an excellent review of all these factors and their possible impact on the future of agriculture, see Rosamond Naylor, "Energy and Resource Constraints on Intensive Agricultural Production," *Annual Reviews of Energy and Environment* 21 (1996): 99–123.

21. Janet McConnaughey, "Scientists Seek Ways to Bring Marine Life Back to World's 'Dead Zones,'" *Los Angeles Times*, August 8, 1999.

22. See, for example, Michael J. Dover and Lee M. Talbot, *To Feed the Earth: Agro-Ecology for Sustainable Development* (Washington, DC: WRI, 1987).

23. The literature on "organic," "low-input," or "ecological" agriculture is enormous. For worldwide examples visit the International Federation of Organic Agricultural Movements at www.ifoam.org/.

24. David Tilman, "The Greening of the Green Revolution," *Nature* 396 (November 19, 1998): 211; see also L. E. Drinkwater, P. Wagoner, and M. Sarrantonio, "Legume-Based Cropping Systems Have Reduced Carbon and Nitrogen Losses," *Nature* 396 (November 19, 1998): 262.

25. *FoodReview* No. 24-1. (Washington, DC : Food and Rural Economics Division, US Department of Agriculture, July 2001)

26. See D. H. Meadows, "Poor Monsanto," in *Whole Earth Review,* Summer 1999, 104.

27. Sandra Postel, Gretchen C. Daily, and Paul R. Ehrlich, "Human Appropriation of Renewable Fresh Water," *Science* 271 (February 9 1996):785–788. This publication is the source for all the numbers that go into figure 3-5.

28. The total capacity of human-made reservoirs is about 5,500 cubic kilometers, but only a bit more than half of that is actually available as sustainable flow.

29. Global capacity for water desalination in 1996 was 6.5 cubic kilometers per year, about 0.1 percent of all human water use. Desalination is very capital- and energy-intensive. Seven of the 10 top countries in desalination capacity were in the Persian Gulf, where other freshwater sources are scarce but nonrenewable fossil fuel energy is cheap. Peter H. Gleick, *The World's Water 1998–99* (Washington, DC: Island Press, 1999), 30.

30. The limit could be and probably will be raised more by further dam building, but the most accessible and largest dam sites have largely been developed. There is an increasing backlash against dams because of their impact on farmland, human settlements, and wildlife. See the final report of the World Commission on Dams (www.dams.org) called *Dams and Development: A New Framework for Decision-Making* (London: Earthscan, 2000).

31. WRI, *World Resources 1998–99*, 188.

32. Gleick, *Water*, 14.

33. Ibid., 1–2.

34. United Nations Development Program, *Human Development Report 1998* (New York: Oxford University Press, 1998), 210.

35. Gleick, *Water*, 2.

36. UN Comprehensive Assessment of the Freshwater Resources of the World, 1997.

37. These examples and many more can be found in Sandra Postel, *Pillar of Sand: Can the Irrigation Miracle Last?* (New York: W. W. Norton, 1999).

38. Lester R. Brown, "Water Deficits Growing in Many Countries," *Eco-Economy Update* (Washington, DC: Earth Policy Institute, August 6, 2002), 2–3.

39. For some case studies see Malin Falkenmark, "Fresh Waters as a Factor in Strategic Policy and Action," in *Global Resources and International Conflict*, edited by Arthur H. Westing (Oxford: Oxford University Press, 1986).

40. The following examples and numbers are taken from Postel, *Pillar*, and from Paul Hawken, Amory Lovins, and Hunter Lovins, *Natural Capital* (New York: Little, Brown, 1999), chapter 11.

41. Very different numbers are being used by different authors for the area of the world's forests. This is both because there exist various definitions of what constitutes a forest, and because the main supplier of data, the FAO, changed its definitions in its year-2000 assessment. In this section we use the new FAO numbers, taken from Forest Resource Assessment (FRA) (Rome: FAO, 2000), www.fao.org/forestry/index.jsp.

42. Dirk Bryant, Daniel Nielsen, and Laura Tangley, *The Last Frontier Forests: Ecosystems and Economies on the Edge* (Washington, DC: WRI, 1997), 1, 9, 12.

43. This estimate is from the UNEP's World Conservation Monitoring Center in the UK (www.unep-wcmc.org/forest/world), including forest in IUCN Conservation Categories I–VI, and is a global average. The protected fraction is roughly the same in the temperate and boreal (northern forests) as in the tropical (southern) forests. Measured as a fraction of the original forest cover—that is, the forested acreage before human deforestation—the percentage should be halved.

44. See Nels Johnson and Bruce Cabarle, "Surviving the Cut: Natural Forest Management in the Humid Tropics" (Washington, DC: WRI, 1993).

45. WCFSD, *Our Forests*, 48.

46. FAO, Provisional Outlook for Global Forest Products Consumption, Production, and Trade to 2010 (Rome: FAO, 1997).

47. Janet N. Abramovitz and Ashley T. Mattoon, "Reorienting the Forest Products Economy," in Brown et al., *State of the World 1999*, 73.

48. Brown et al., *State of the World 1999*, 65.

49. Abramovitz and Mattoon, "Forest Products," 64.

50. World Resources 1998-99: Environmental change and human health (Washington, DC, World Resources Institute, 1998).

51. This list is adapted from Gretchen C. Daily, editor, *Nature's Services: Societal Dependence on Natural Ecosystems* (Washington, DC: Island Press, 1997), 3–4.

52. See Robert Costanza et al., "The Value of the World's Ecosystem Services and Natural Capital," *Nature* 387 (1997): 253–260. Costanza and colleagues estimated (conservatively) the value of natural services at $33 trillion per year, at a time when the gross world economic product was $18 trillion per year.

53. Robert M. May, "How Many Species Inhabit the Earth?" *Scientific American*, October 1992, 42.

54. Joby Warrick, "Mass Extinction Underway, Majority of Biologists Say," *Washington Post*, April 21, 1998, A4.

55. Don Hinrichson, "Coral Reefs in Crisis," *Bioscience,* October 1997.

56. See, for example, "Extinction: Are Ecologists Crying Wolf?" *Science* 253 (August 16, 1991): 736, and other articles in the same issue, which express the serious concerns of ecologists.

57. Species Survival Commission, 2000 IUCN Red List of Threatened Species (Gland, Switzerland: International Union for the Conservation of Nature, 2000), as quoted in Brown, "Water Deficits," 69.

58. Constance Holden, "Red Alert for Plants," *Science* 280 (April 17, 1998): 385.

59. SSC, IUCN Red List, 1.

60. WWF, *Living Planet Report 2002.*

61. "World Scientists' Warning to Humanity," December 1992, signed by more than 1,600 scientists, including 102 Nobel laureates, available from Union of Concerned Scientists, 26 Church Street, Cambridge, MA 02238.

62. *Commercial energy* refers to that sold in a market; it does not count the energy used by people who gather wood, dung, and other biomass for their own use. *Noncommercial energy sources* are mostly renewable, but are not necessarily harvested sustainably. They are estimated to constitute around 7 percent of total energy consumption. WRI, *World Resources 1998–99*, 332.

63. U.S. Energy Information Administration, *International Energy Outlook 2003*, table A1, "World Total Energy Consumption by Region, Reference Case, 1990–2025 (Quadrillion BTU)," www.eia.doe.gov/oiaf/ieo/.

64. International Energy Agency, *World Energy Outlook 2002* (Vienna: IEA, 2002), www.worldenergyoutlook.org/weo/pubs/weo2002/weo2002.asp. Longer-term scenarios can be found in World Energy Council, "Global Energy Scenarios to 2050 and Beyond," 1999, www.worldenergy.org/wec-geis/edc/.

65. Bent Sørensen, "Long-Term Scenarios for Global Energy Demand and Supply," Energy & Environment Group, Roskilde University, January 1999.

66. *Production* is a misleading word for the process of taking fossil fuels out of the ground. Nature is the producer of these fuels, over millions of years. Human beings do not "produce" them; they extract, exploit, harvest, pump, mine, or take them. However, *production* is the word commonly used, especially in such terms as *reserve–production ratio,* so we have used it as well.

67. Of course the capital plants for discovery, mining, pumping, transporting, and refining also burn fuels. If there were no other limits, the ultimate limit to the use of fossil fuels would come at the point where it takes as much energy to get them as they contain. See Charles A. S. Hall and Cutler J. Cleveland, "Petroleum Drilling and Production in the United States: Yield per Effort and Net Energy Analysis," *Science* 211 (February 6, 1981): 576.

68. This information and most of the data we quote on this topic come from Amory Lovins and the Rocky Mountain Institute. For more detailed information on energy efficiency options in transportation, industry, and buildings, see *Scientific American* 263, no. 3 (September 1990).

69. UNDP, *Human Development Indicators 2003*, http://hdr.undp.org/reports/global/2003/indicator/index.html.

70. Total current human use of fossil fuel constitutes a flow of power equal to about 5 terawatts (billion kilowatts). The constant inflow of the sun to the earth's surface is 80,000 terawatts.

71. Lester Brown et al., *Vital Signs 2000* (New York: W. W. Norton, 2000), 58. Both figures are in 1998 dollars.

72. American Wind Energy Association, "Record Growth for Global Wind Power in 2002" (Washington, DC: AWEA, March 3, 2002), 1.

73. Peter Bijur, Global Energy Address to the 17th Congress of the World Energy Council, Houston, September 14, 1998.

74. The most promising storage mechanism may be hydrogen made from solar-electric splitting of water molecules. Hydrogen may also be the answer to vehicle propulsion in the future. For a review, see chapter 5 in Brown, *Eco-Economy*.

75. For a systematic examination of these possibilities, see John E. Tilton, editor, *World Metal Demand* (Washington, DC: Resources for the Future, 1990).

76. Organization for Economic Cooperation and Development, *Sustainable Development: Critical Issues* (Paris: OECD, 2001), 278.

77. Personal communication from Aleksander Mortensen in the Norwegian recycling firm Tomra ASA (www.tomra.no). In 2001 the world primary aluminium production was some 21 million tons. In addition some 2.2 million tons of aluminum scrap were recovered (www.world-aluminum.org/iai/stats/index.asp). Information on beverage containers is from www.canadean.com; on recycling, from www.container-recycling.org.

78. WRI, *Resource Flows: The Material Basis of Industrial Economies* (Washington, DC: WRI, 1997), gives a summary of the declining materials intensity in four industrial economies.

79. For an overview of the production of waste in various nations, see OECD, *Environmental Data: Compendium 1999* (Paris: OECD ,1999).

80. Earl Cook, "Limits to Exploitation of Nonrenewable Resources," *Science* 20 (February 1976).

81. International Institute for Environment and Development and World Business Council for Sustainable Development, *Breaking New Ground: Mining, Minerals, and Sustainable Development* (London: Earthscan, 2002), 83.

82. The United States, Japan, Great Britain, France, Germany, Italy, and Canada.

83. The information in the preceding section is taken from Urs Weber, "The Miracle of the Rhine," *UNESCO Courier* (June 2000), and from the database of the Web site of the International Commission for the Protection of the Rhine, www.iksr.org.

84. Bjørn Lomborg, *The Skeptical Environmentalist: Measuring the Real State of the World* (Cambridge: Cambridge University Press, 2001), 203.

85. Ibid., 167–176.

86. Ibid., 205.

87. WCED, *Our Common Future*, 224.

88. Robert T. Watson, chair, Intergovernmental Panel on Climate Change, presenting the key conclusions of the IPCC Third Assessment Report (Climate Change 2001) to the Sixth Conference of Parties to the United Nations Framework Convention on Climate Change, July 19, 2001. Available at www.ipcc.ch.

89. D. H. Meadows et al., *Limits to Growth* (New York: Universe Books, 1972), 79.

90. WWF, *Living Planet Report 1999* (Gland, Switzerland: WWF, 1999), 8.

91. R. T. Watson et al., *Climate Change 2001: Synthesis Report, Intergovernmental Panel on Climate Change* (Geneva, Switzerland: IPCC, 2001). Also available along with numerous illustrations at www.ipcc.ch.

92. For a colorful presentation of the skeptic's view on climate and all other environmental issues, see Lomborg, *Environmentalist.*

93. See the vastly informative Web site of the Climatic Research Unit at the University of East Anglia, Norwich, UK, www.cru.uea.ac.uk.

94. See, for example, "Global Warming. Stormy Weather," *Time,* November 13, 2000, 35–40, with regional weather forecasts for Europe to 2050.

95. Watson et al., *Climate Change 2001.*

96. These data come from ice cores drilled deep into the Antarctic ice sheet. The polar ice has accumulated over thousands of years, layer after layer, and in each layer are trapped tiny air bubbles, preserved from prehistoric times. Isotopic analysis can date the core layers and provide clues to past temperatures; direct analyses of the air bubbles give the carbon dioxide and methane concentrations.

97. Committee on Abrupt Climate Change, *Abrupt Climate Change—Inevitable Surprises* (Washington, DC: National Academy Press, 2002), 1.

98. These promising avenues are explored in depth in Ernst von Weizsäcker, Amory Lovins, and L. Hunter Lovins, *Factor Four: Doubling Wealth, Halving Resource Use* (London: Earthscan, 1997).

99. UNEP, *Global Environmental Outlook 2000* (London: Earthscan, 1999).

100. We have adapted this formulation from one originally put forward by Amory Lovins.

Chapter 4. World3: The Dynamics of Growth in a Finite World

1. Isaac Asimov, *Prelude to Foundation* (New York: Doubleday, 1988), 10.

2. An example of this approach is provided by Wolfgang Lutz, editor, *The Future Population of the World: What Can We Assume Today?* revised and updated edition (London: Earthscan, 1996).

3. The CD contains a STELLA© flow diagram of World3, the full model for Scenario 1, and an interface with which you can reproduce and examine the details of all 11 scenarios published in this book. To obtain ordering information, consult www.chelseagreen.com.

4. The concept of carrying capacity was originally defined for relatively simple population–resource systems. For example, it was used to talk about the number of cattle that could be maintained on a specific pasture without degrading the land. For human populations the term *carrying capacity* is much more complex, and there is no universally accepted definition. It is complex because people take many types of resources from the environment; they generate many kinds of wastes; their impact on the environment is affected by a great variety of technologies, institutions, and lifestyles. There is disagreement about the minimum time a system must be able to persist in order to be considered sustainable. Nor is there agreement about how to allow for the demands of other species. In any event, carrying capacity is a dynamic concept. It is always changing with weather, technological advance, consumption patterns, climate, and other factors. We use the term loosely to designate the number of people, at prevailing circumstances, that could be supported on the globe for a long time—at least many decades—without deteriorating the

overall productivity of the planet. See Joel E. Cohen, *How Many People Can the Earth Support?* (New York: W. W. Norton, 1995).

5. Other authors have found this categorization to be useful for thinking about the future. See, for example, William R. Caton, *Overshoot: The Ecological Basis of Revolutionary Change* (Chicago: University of Illinois Press, 1982), 251–254.

6. M. Wackernagel et al., "Ecological Footprints of Nations: How Much Nature Do They Use? How Much Nature Do They Have?" (Xalapa, Mexico: Centro de Estudios para la Sustentabilidad [Center for Sustainability Studies], March 10, 1997).

7. Only in Scenarios 0 and 1 do we assume that the initial endowment of nonrenewable resources is half that amount.

8. There are 209 of these chemicals, all created by adding chlorine atoms in various positions to the 2 joined benzene rings of the molecule called biphenyl. They are human-synthesized, not normally found in nature.

9. Sören Jensen, *New Scientist* 32 (1966): 612.

10. For a popular and comprehensive account of the endocrine disruptor story, see Theo Colborn, Dianne Dumanoski, and John P. Myers, *Our Stolen Future* (New York: Dutton, 1996), which also contains hundreds of references to the rapidly growing scientific literature on this subject.

11. The Soviet Union stopped making PCBs only in 1990.

12. J. M. Marquenie and P. J. H. Reijnders, "Global Impact of PCBs with Special Reference to the Arctic," Proceedings of the 8th International Congress of Comite Arctique Internationale, Oslo, September 18–22, 1989 (Lillestrom, Norway: NILU).

13. A. Larson, "Pesticides in Washington State's Ground Water, A Summary Report, 1988–1995," Report 96-303, Washington State Pesticide Monitoring Program, January 1996.

14. See "New Cause of Concern on Global Warming," *New York Times*, February 12, 1991.

15. W. M. Stigliani, "Chemical Time Bombs," *Options* (Laxenburg, Austria: International Institute of Applied Systems Analysis, September 1991), 9.

16. In addition to negotiations and research on the destruction of the ozone layer, described in chapter 5, and on global climate change, mentioned in chapter 3, there are major international research programs on "global change" sponsored by the International Council of Scientific Unions (ICSU) and the World Meteorological Organization (WMO). They include the International Geosphere-Biosphere Program (IGBP), the World Climate Research Program (WCRP), and the International Human Dimensions Program (IHDP). There are also numerous national and regional efforts, such as the U.S. Global Change Research Program.

17. The term *consumer goods per capita* represents the fraction of industrial output that is directed to consumer goods such as cars, appliances, and clothing. It is roughly 40 percent of total output. It does not include food, services, or investment, which are calculated separately. In the model, consumer goods, industrial output, and services represent real, physical things, but they are measured in dollars, because that is the only metric used in the economic data. In the original model we calibrated everything to 1968 dollars and have seen no reason to change that, since we are mainly interested in relative, not absolute welfare measures. Since it is hard for people decades later to relate to measures in 1968 dollars (worth about four times more than year-2000 dollars), we limit our discussion in this book to relative economic terms.

Chapter 5. Back from Beyond the Limits: The Ozone Story

1. A variety of chlorine- and bromine-containing chemicals have the capacity to erode the stratospheric ozone layer—methyl bromide, a soil fumigant; carbon tetrachloride, a cleaning solvent; halons used in firefighting; and others. But the greatest threat comes from CFCs, a family of compounds containing fluorine, hydrogen, and chlorine. These have been the subject of the most research, and they are the focus of most international control efforts. So we will focus our story on them.

2. Arjun Makhijani, Annie Makhijani, and Amanda Bickel, *Saving Our Skins: Technical Potential and Policies for the Elimination of Ozone-Depleting Chlorine Compounds* (Washington, DC: Environmental Policy Institute and the Institute for Energy and Environmental Research, September 1988), 83. Available from the Environmental Policy Institute, 218 O Street SE, Washington, DC 20003.

3. Ibid., 77.

4. B. K. Armstrong and A. Kricker, "Epidemiology of Sun Exposure and Skin Cancer," *Cancer Surveys* 26 (1996): 133-153.

5. See, for example, Robin Russell Jones, "Ozone Depletion and Cancer Risk," *Lancet* (August 22, 1987), 443; "Skin Cancer in Australia," *Medical Journal of Australia* (May 1, 1989); Alan Atwood, "The Great Cover-up," *Time* (Australia), 27 February 1989; Medwin M. Mintzis, "Skin Cancer: The Price for a Depleted Ozone Layer," *EPA Journal* (December 1986).

6. Osmund Holm-Hansen, E. W. Heibling, and Dan Lubin, "Ultraviolet Radiation in Antarctica: Inhibition of Primary Production," *Photochemistry and Photobiology* 58, no. 4 (1993): 567–570.

7. A. H. Teramura and J. H. Sullivan, "How Increased Solar Ultraviolet-B Radiation May Impact Agricultural Productivity," in *Coping with Climate Change* (Washington, DC: Climate Institute, 1989), 203.

8. Richard S. Stolarski and Ralph J. Cicerone, "Stratospheric Chlorine: A Possible Sink for Ozone," *Canadian Journal of Chemistry* 52 (1974): 1610.

9. Mario J. Molina and F. Sherwood Rowland, "Stratospheric Sink for Chlorofluoromethanes: Chlorine Atomic Catalysed Destruction of Ozone," *Nature* 249 (1974): 810. For this research Molina and Rowland were awarded the Nobel Prize in Chemistry in 1995.

10. Quoted in Richard E. Benedick, *Ozone Diplomacy* (Cambridge, MA: Harvard University Press, 1991), 12.

11. J. C. Farman, B. G. Gardiner, and J. D. Shanklin, "Large Losses of Total Ozone in Antarctica Reveal Seasonal ClO/NO_2 Interaction," *Nature* 315 (1985): 207.

12. The period during which scientists were seeing low ozone readings and yet not "seeing" them is described well in Paul Brodeur, *Annals of Chemistry,* 71.

13. J. G. Anderson, W. H. Brune, and M. J. Proffitt, "Ozone Destruction by Chlorine Radicals within the Antarctic Vortex: The Spatial and Temporal Evolution of $ClO-O_3$ Anticorrelation Based on in Situ ER-2 Data," *Journal of Geophysical Research* 94 (August 30, 1989): 11, 474.

14. Mario J. Molina, "The Antarctic Ozone Hole," *Oceanus* 31 (Summer 1988).

15. DuPont dropped its search for CFC substitutes upon the election of Ronald Reagan as president in 1980.

16. The political process is described clearly and fully by Richard Benedick, who was the chief negotiator for the United States, in R. E. Benedick, *Ozone Diplomacy: New Directions in Safeguarding the Planet*, 2nd ed. (Cambridge, MA, and London: Harvard University Press, 1998).

17. Ibid., 215.

18. United Nations Environment Program, "Synthesis of the Reports of the Scientific Assessment Panel and Technology and Economic Assessment Panel on the Impact of HCFC and Methyl Bromide Emissions," Nairobi, March 1995, section 4.

19. World Meteorological Organization, "Scientific Assessment of Ozone Depletion: 2002," *Global Ozone Research and Monitoring Project Report 47*, available at www.unep.org/ozone.

20. By then the UNEP office that collects this information had stopped producing aggregated time series data because of the variation in reporting quality from year to year. See "Production and Consumption of Ozone Depleting Substances under the Montreal Protocol 1989–2000" (Nairobi: UNEP, 2002), available at www.unep.ch/ozone/. Production statistics are in tables 1 and 2, 18 onward.

21. F. A. Vogelsberg, "An Industry Perspective: Lessons Learned and the Cost of the CFC Phaseout," paper presented at the International Conference on Ozone Protection Technologies, Washington, DC, October 1996.

22. Richard A. Kerr, "Deep Chill Triggers Record Ozone Hole," *Science* 282 (October 16, 1998): 391.

23. WMO, "Scientific Assessment," xiv and xv.

24. World Resources Institute, *World Resources 1998–99* (New York: Oxford University Press, 1998), 178. See also Tim Beardsley, "Hot Coolants," *Scientific American,* July 1998, 32.

25. Mario J. Molina, "Stratospheric Ozone: Current Concerns," paper presented at the Symposium on Global Environmental Chemistry—Challenges and Initiatives, 198th National Meeting of the American Chemical Society, September 10–15, 1989, Miami Beach, Florida.

26. The Industrial Coalition for Ozone Layer Protection, 1440 New York Avenue NW, Suite 300, Washington, DC 20005.

27. WMO, "Scientific Assessment," xxxix.

Chapter 6. Technology, Markets, and Overshoot

1. But it is of course true that if you assume sufficiently rapid technological advance, and instantaneous implementation of the resulting new technologies, all problems associated with a growing ecological footprint can be solved. We described the changes to achieve such advances in Scenario 0, "Infinity In, Infinity Out," in chapter 4.

2. Markets have their own temporary overshoots and undershoots, which we have modeled in many other contexts, but for simplicity we left short-term price instabilities out of World3; they are not strongly linked to global changes that extend over many decades.

3. We took that line about control with nature as instrument from one of the most wonderful essays on technology ever written: C. S. Lewis, "The Abolition of Man," in Herman Daly, *Toward a Steady-State Economy* (San Francisco: Freeman Press, 1973).

4. That assumption was made in 1970, and at that time we implemented those technologies as discontinuous steps in the simulated year 1975. By the real year 1990 some of the technologies had begun to be incorporated structurally into the world economy. Therefore we made some permanent adjustments to the numbers within World3—for instance, significantly reducing resource use per unit of industrial output. These numerical changes are explained in detail in the appendix to Donella H. Meadows, Dennis L. Meadows, and Jorgen Randers, *Beyond the Limits* (Post Mills, VT: Chelsea Green Publishing Company, 1992).

5. We used this "adaptive technologies" formulation already in the early 1970s in the technical report on the *Limits to Growth* study. See Dennis L. Meadows et al., *Dynamics of Growth in a Finite World* (Cambridge, MA: Wright-Allen Press, 1974), 525–537.

6. Lester Brown et al., *Vital Signs 2000* (New York: W. W. Norton, 2000), 53.

7. Brown et al., *Vital Signs 2000*, 41.

8. United Nations Food and Agriculture Organization, "The State of World Fisheries and Aquaculture 2002," www.fao.org/docrep/005/y7300e/y7300e00.htm.

9. Lester Brown, *Eco-Economy* (New York: W. W. Norton, 2001), 51–55.

10. Fact sheets of the World Wide Fund for Nature Endangered Seas Campaign, 2003, www.panda.org/campaigns/marine/sturgeon.

11. The classic analysis of this phenomenon is Garrett Hardin's "The Tragedy of the Commons," *Science*, 162(1968):1243–1248.

12. *Audubon* (September–October 1991), 34.

13. *Dagens Naeringsliv* (Norwegain business journal), Oslo (December 9, 2002), 10.

14. Japanese journalist to Paul Ehrlich, in *Animal Extinctions: What Everyone Should Know*, edited by R. J. Hoage (Washington, DC: Smithsonian Institution Press, 1985), 163.

15. Erling Moxness, "Not Only the Tragedy of the Commons: Misperceptions of Feedback and Policies for Sustainable Development," *System Dynamics Review* 16, no. 4 (Winter 2000): 325–348.

Chapter 7. Transitions to a Sustainable System

1. See Duane Elgin, *Voluntary Simplicity*, revised edition (New York: Quill, 1998), as well as Joe Dominguez and Vicki Robin, *Your Money or Your Life: Transforming Your Relationship with Money and Achieving Financial Independence* (New York: Penguin USA, 1999).

2. World Commission on Environment and Development, *Our Common Future* (Oxford: Oxford University Press, 1987).

3. Herman Daly is one of the few people who have begun to think through what kinds of economic institutions might work to maintain a desirable sustainable state. He comes up with a thought-provoking mixture of market and regulatory devices. See, for example, Herman Daly, "Institutions for a Steady-State Economy," in *Steady State Economics* (Washington, DC: Island Press, 1991).

4. Aurelio Peccei, *The Human Quality* (New York: Pergamon Press, 1977), 85.

5. John Stuart Mill, *Principles of Political Economy*, (London: John W Parker, West Strand, 1848).

6. A good example is the biannual WWF *Living Planet Report* published by World Wide Fund for Nature International, Gland, Switzerland, which provides data on trends in global biodiversity and the ecological footprint of nations.

7. See Paul Hawken, Amory Lovins, and L. Hunter Lovins, *Natural Capitalism* (Boston: Back Bay Books, 2000).

8. Lewis Mumford, *The Condition of Man* (New York: Harcourt Brace Jovanovich, 1944), 398–399.

Chapter 8. Tools for the Transition to Sustainability

1. Donald Worster, editor, *The Ends of the Earth* (Cambridge: Cambridge University Press, 1988), 11–12.

2. Ralph Waldo Emerson, Lecture on "War," delivered in Boston, March 1838. Reprinted in *Emerson's Complete Works*, vol. 11 (Boston: Houghton Mifflin, 1887), 177.

3. Examples of networks known to the authors and in their field of interest are the Balaton Group (www.unh.edu/ipssr/Balaton.html), Northeast Organic Farming Association (NOFA), Center for a New American Dream (CNAD; www.newdream.org), Greenlist (www.peacestore.us/Public/Greenlist), Greenclips (www.greenclips.com), Northern Forest Alliance (www.northernforestalliance.org), Land Trust Alliance (www.lta.org), International Simulation and Gaming Association (ISAGA; www.isaga.info), and Leadership for Environment and Development (LEAD).

4. Such an intermediate step is illustrated by ICLEI, an international association of (currently 450) local governments implementing sustainable development. See www.iclei.org.

5. R. Buckminster Fuller, *Critical Path* (New York: St. Martin's Press, 1981).

6. Abraham Maslow, *The Farthest Reaches of Human Nature* (New York: Viking Press, 1971).

7. J. M. Keynes, foreword to *Essays in Persuasion* (New York: Harcourt Brace, 1932).

8. Aurelio Peccei, *One Hundred Pages for the Future* (New York: Pergamon Press, 1981), 184–185.

Appendix 1. Changes from World3 to World3-03

1. Dennis L. Meadows et al., *Dynamics of Growth in a Finite World* (Cambridge, MA: Wright-Allen Press, 1974).

2. Donella H. Meadows, Dennis L. Meadows, and Jorgen Randers, *Beyond the Limits*, (Post Mills, VT: Chelsea Green Publishing Company, 1992).

3. To obtain ordering information, consult www.chelseagreen.com.

Appendix 2. Indicators of Human Welfare and Ecological Footprint

1. Jay W. Forrester, *World Dynamics* (Cambridge, MA: Wright-Allen Press, 1971).
2. United Nations Development Program, *Human Development Report 2001* (New York and Oxford: Oxford University Press, 2001).
3. Ibid., 240.
4. The details of the calculation of the HDI are presented in ibid., 239–240.
5. UNDP, *Human Development Report 2000* (New York and Oxford: Oxford University Press, 2000), 144.
6. Mathis Wackernagel et al., "National Natural Capital Accounting with the Ecological Footprint Concept," *Ecological Economics* 29 (1999): 375–390.
7. Mathis Wackernagel et al., "Tracking the Ecological Overshoot of the Human Economy," *Proceedings of the Academy of Science* 99, no. 14 (Washington, DC, 2002): 9266–9271. See also figure P-1 in the authors' preface to the present book.
8. World Wide Fund for Nature, *Living Planet Report 2002* (Gland, Switzerland: WWF, 2002).
9. More details of the calculation of the ecological footprint are presented in ibid., 30.

List of Tables and Figures with Sources

Authors' Preface

Figure P-1 Ecological Footprint versus Carrying Capacity
  Mathis Wackernagel et al., "Tracking the Ecological Overshoot of the Human
    Economy", *Proceedings of the Academy of Science* 99, no. 14 (2002): 9266–9271,
    www.pnas.org/cgi/doi/10.1073/pnas.142033699.

Chapter 1

Figure 1-1 World Population
  World Population Data Sheet (Washington, DC: Population Reference Bureau)
    http://www.prb.org (accessed in various years).
  World Population Prospects as Assessed in 1994 (New York: United Nations, 1994).
  Donald J. Bogue, *Principles of Demography* (New York: John Wiley and Sons, 1969).

Figure 1-2 World Industrial Production
  Statistical Yearbook (New York: United Nations, various years).
  Demographic Yearbook (New York: United Nations, various years).
  World Population Data Sheet, (Washington, DC: Population Reference Bureau)
    http://www.prb.org (accessed in various years).
  Industrial Statistical Yearbook (New York: United Nations, various years).
  Monthly Bulletin of Statistics (New York: United Nations, various dates).

Figure 1-3 Carbon Dioxide Concentration in the Atmosphere
  C. D. Keeling and T. P. Whorf, "Atmospheric CO_2 Concentrations (ppmv) Derived
    from *In Situ* Air Samples Collected at Mauna Loa Observatory, Hawaii," *Trends:*

A Compendium of Data on Global Change, (August 13, 2001) http://cdiac.esd.ornl.gov/trends/.

A. Neftel, H. Friedli, E. Moor, H. Lötscher, H. Oeschger, U. Siegenthaler, and B. Stauffer. 1994. "Historical CO_2 Record from the Siple Station Ice Core," *Trends: A Compendium of Data on Global Change* (1994) http://cdiac.esd.ornl.gov/trends/co2/siple.htm.

Table 1–1 Worldwide Growth in Selected Human Activities and Products 1950–2000
CRB Commodity Yearbook (New York: Commodity Research Bureau, various years).

International Petroleum Monthly (Washington, D.C.: Energy Information Administration, U.S. Dept. of Energy) http://www.eia.doe.gov/ipm (accessed 1/30/2002).

International Energy Outlook 1998 (Washington, D.C.: Energy Information Administration, U.S. Dept. of Energy, 1998) http://www.eia.doe.gov/oiaf/ieo/.

International Energy Annual 1999 (Washington, D.C.: Energy Information Administration, U.S. Dept. of Energy, 1999) http://www.eia.doe.gov/iea/.

Ward's Motor Vehicle Facts and Figures 2000 (Southfield, MI: Ward's Communications, 2000).

UN Food and Agriculture Organization FAOSTAT on-line database, http://apps.fao.org/.

World Population Data Sheet, (Washington, DC: Population Reference Bureau) http://www.prb.org (accessed in various years).

Energy Statistics Yearbook (New York: United Nations, various years).

Statistical Yearbook (New York: United Nations, various years).

World Motor Vehicle Data, 1998 (Detroit: Automobile Manufacturers Association, 1998).

World Population Prospects as Assessed in 1994 (New York: United Nations, 1994).

Figure 1-4 Alternative Scenarios for Global Population and Human Welfare

Chapter 2

Figure 2-1 World Soybean Production
  Lester R. Brown et al., *Vital Signs 2000: the Environmental Trends That are Shaping Our Future* (New York : W.W. Norton, 2000).
  UN Food and Agriculture Organization FAOSTAT on-line database, http://apps.fao.org/.

Figure 2-2 World Urban Population
  World Urbanization Prospects: the 1999 Revision (New York: United Nations, 2001).

Figure 2-3 Linear versus Exponential Growth of Savings

Table 2–1 Doubling Times

Table 2–2 Nigeria's Population Growth, Extrapolated
  U.S. Census Bureau International Data Base, http://www.census.gov/ipc/www/idbnew.html.

Figure 2-4 World Demographic Transition
  The World Population Situation in 1970 (New York: United Nations, 1971).
  World Population Prospects: the 2000 Revision (New York: United Nations, 2001) http://www.un.org/popin/.

Table 2–3 Additions to World Population
  The World Population Situation in 1970 (New York: United Nations, 1971).
  World Population Prospects: the 2000 Revision (New York: United Nations, 2001) http://www.un.org/popin/.

Figure 2-5 World Annual Population Increase
  World Population Prospects 2000 (New York: United Nations, 2000).
  Donald J. Bogue, *Principles of Demography* (New York: John Wiley and Sons, 1969).

Figure 2-6 Demographic Transitions in Industrialized Countries and in Less Industrialized Countries

Nathan Keyfitz and W. Flieger, *World Population: an Analysis of Vital Data* (Chicago: Univ. Chicago Press, 1968).

J. Chesnais, *The Demographic Transition: Stages, Patterns, and Economic Implications; a Longitudinal Study of Sixty-Seven Countries Covering the Period 1720–1984* (New York Oxford University Press, 1992).

Demographic Yearbook (New York: United Nations, various years).

World Population Data Sheet (Washington, DC: Population Reference Bureau) http://www.prb.org (accessed in various years).

United Kingdom Office of Population Censuses & Surveys, *Population Trends*, no. 52 (London: HMSO, June 1988).

United Kingdom Office for National Statistics (ONS), *National Statistics Online: Birth Statistics: Births and patterns of family building England and Wales (FM1)*, http://www.statistics.gov.uk/STATBASE/Product.asp?vlnk=5768.

Statistical Yearbook of the Republic of China (Taipei: Directorate-General of Budget, Accounting & Statistics, Executive Yuan, Republic of China, 1995).

Figure 2-7 Birth Rates and Gross National Income per Capita in 2001

World Population Data Sheet 2001 (Washington, DC: Population Reference Bureau, 2001) http://www.prb.org.

World Bank, "World Development Indicators (WDI) Database," http://www.worldbank.org/data/dataquery.html (accessed 1/15/04).

Figure 2-8 Flows of Physical Capital in the Economy of World3

Figure 2-9 U.S. Gross National Income by Sector

U.S. Dept. of Commerce, *Bureau of Economic Analysis Interactive Access to National Income and Product Accounts Tables*, http://www.bea.doc.gov/bea/dn/nipaweb/.

Figure 2-10 Per Capita GNI of the Top 10 Most Populous Countries and the European Monetary Union

World Development Indicators CD-ROM (Washington, DC: World Bank, 2002).

Figure 2-11 Global Disparities

World Development Indicators CD-ROM (Washington, DC: World Bank, 1999).

UN Food and Agriculture Organization FAOSTAT on-line database, http://apps.fao.org/ (accessed 2/27/02).

Figure 3-5 Freshwater Resources

Peter Gleick, *The World's Water 2000–2001: the Biennial Report on Freshwater Resources* (Washington, DC: Island Press, 2000).

S. L. Postel, G. C. Daly, P. R. Erlich, "Human Appropriation of Renewable Fresh Water," *Science* 271 (Feb. 9 1996):785-788.

Donald J. Bogue, *Principles of Demography* (New York: John Wiley and Sons, 1969).

World Population Prospects as Assessed in 1994 (New York: United Nations, 1994).

World Population Prospects as Assessed in 2000 (New York: United Nations, 2000).

Figure 3-6 U.S. Water Use

Peter H. Gleick, *The World's Water* (Washington, DC: Island Press, 1998).

Peter Gleick, *The World's Water 2000–2001: the Biennial Report on Freshwater Resources* (Washington, DC: Island Press, 2000).

Figure 3-7 Remaining Frontier Forests

The Last Frontier Forests: Ecosystems and Economies on the Edge (World Resources Institute Forest Frontiers Initiative, 1997) http://www.wri.org/ffi/lff-eng/.

Figure 3-8 Some Possible Paths of Tropical Deforestation

Figure 3-9 World Wood Use

UN Food and Agriculture Organization FAOSTAT on-line database, http://apps.fao.org/.

Figure 3-10 World Energy Use

Energy Statistics Yearbook (New York: United Nations, various years).

U.S. Dept. of Energy, Energy Information Administration International Energy Data on-line database, http://www.eia.doe.gov/emeu/international/energy.html.

International Energy Outlook 2001 (Washington, D.C.: Energy Information Administration, U.S. Dept. of Energy, 2001) http://www.eia.doe.gov/oiaf/ieo/.

for Integrated Environmental Assessments (HYDE, version 1.1)" (Bilthoven, the Netherlands: National Institute of Public Health and the Environment, 1997).

U.S. Bureau of Mines, *Minerals Yearbook* (Washington, DC: Government Printing Office, various years).

U.S. Geological Survey, Statistical Compendium on-line resource, http://minerals.usgs.gov/minerals/pubs/stat/.

CRB Commodity Yearbook (New York: Commodity Research Bureau, various years).

Figure 3-17 World Consumption of Steel

C. G. M. Klein Goldewijk and J. J. Battjes, "A Hundred Year (1890-1990) Database for Integrated Environmental Assessments (HYDE, version 1.1)" (Bilthoven, the Netherlands: National Institute of Public Health and the Environment, 1997).

U.S. Bureau of Mines, Minerals Yearbook (Washington, DC: Government Printing Office, various years).

U.S. Geological Survey, Statistical Compendium on-line resource, http://minerals.usgs.gov/minerals/pubs/stat/.

CRB Commodity Yearbook (New York: Commodity Research Bureau, various years).

Table 3-2 Life Expectancies of Identified Reserves for Eight Metals

Mining, Minerals and Sustainable Development Project (MMSD), *Breaking New Ground: Mining, Minerals and Sustainable Development* (London: Earthscan, 2002) http://www.iied.org/mmsd/finalreport/.

Figure 3-18 The Declining Quality of Copper Ore Mined in the United States

U.S. Bureau of Mines, *Minerals Yearbook* (Washington, DC: Government Printing Office, various years).

U.S. Geological Survey, Statistical Compendium on-line resource, http://minerals.usgs.gov/minerals/pubs/stat/.

Figure 3-19 Depletion of Mineral Ores Greatly Increases the Mining Wastes Generated in Their Production

Figure 3-20 Decreasing Human and Environmental Contamination

DDT: IVL Swedish Environmental Research Institute, *Swedish Environmental Monitoring Surveys Database*, http://www.ivl.se/miljo/projekt/dvsb/ (accessed December, 2001).

Cesium-137: *AMAP Assessment Report: Arctic Pollution Issues* (Oslo, Norway: Arctic Monitoring and Assessment Programme, 1998) http://www.amap.no/Assessment/ScientificBackground.htm.

Lead: *America's Children and the Environment: Measures of Contaminants, Body Burdens, and Illnesses*, 2nd ed. (Washington, DC: Environmental; Protection Agency, Feb. 2003) http://www.epa.gov/envirohealth/children/ace_2003.pdf.

Figure 3-21 Trends in Emissions of Selected Air Pollutants

World Development Indicators CD-ROM (Washington, DC: World Bank, 2001).

OECD Environmental Data: Compendium (Paris: Organisation for Economic Co-Operation and Development, various years).

CO_2: G. Marland, T. A. Boden, and R. J. Andres, "Global, Regional, and National Fossil Fuel CO_2 Emissions," *Trends: A Compendium of Data on Global Change*, http://cdiac.esd.ornl.gov/trends/emis/em_cont.htm.

SO_x and NO_x: World Resources Database CD-ROM Electronic Resource (Washington, D.C.: World Resources Institute, 2000).

Energy use: *Energy Balances of Organization for Economic Cooperation and Development (OECD) Countries*, on diskette (Paris: Organisation for Economic Co-Operation and Development, various years).

Figure 3-22 Oxygen Levels in Polluted Waters

Andrew Goudie, *The Human Impact on the Natural Environment* (Oxford: Blackwell, 1993), 224.

P. Kristensen and H. Ole Hansen, *European Rivers and Lakes: Assessment of Their Environmental State* (Copenhagen: European Environmental Agency, 1994), 49.

OECD Environmental Data: Compendium (Paris: Organisation for Economic Co-Operation and Development, 1999), 85.

New York Harbor Water Quality Survey (New York: NY Department of Environmental Protection, 1997), 55.

Bjørn Lomborg, *The Skeptical Environmentalist: Measuring the Real State of the World* (Cambridge: Cambridge University Pres, 2001), 203.

Figure 3-23 Global Greenhouse Gas Concentrations

CFCs: M.A.K. Khalil and R. A. Rasmussen, "Globally Averaged Atmospheric CFC-11 Concentrations: Monthly and Annual Data for the Period 1975–1992," Carbon Dioxide Information Analysis Center (CDIAC), http://cdiac.esd.ornl.gov/ndps/db1010.html.

CH_4: D.M. Etheridge, I. Pearman, P.J. Fraser, "Concentrations of CH_4 from the Law Dome (East Side, "DE08" Site) Ice Core(a)," Carbon Dioxide Information Analysis Center (9/1/1994), http://cdiac.esd.ornl.gov/ftp/trends/methane/lawdome.259.

C. D. Keeling and T. P. Whorf, "Atmospheric CO_2 Concentrations (ppmv) Derived from *In Situ* Air Samples Collected at Mauna Loa Observatory, Hawaii," *Trends: A Compendium of Data on Global Change* (August 13, 2001), http://cdiac.esd.ornl.gov/trends/.

A. Neftel, H. Friedli, E. Moor, H. Lötscher, H. Oeschger, U. Siegenthaler, and B. Stauffer, "Historical CO_2 Record from the Siple Station Ice Core," *Trends: A Compendium of Data on Global Change* (1994)
http://cdiac.esd.ornl.gov/trends/co2/siple.htm.

N_2O: J. Flückiger, A. Dällenbach, B. Stauffer, "N_2O Data Covering the Last Millennium," (1999) NOAA/NGDC Paleoclimatology Program,
http://www.ngdc.noaa.gov/paleo/gripn2o.html.

R. G. Prinn et al., "A History of Chemically and Radiatively Important Gases in Air Deduced from ALE/GAGE/AGAGE" *Journal of Geophysical Research* 115: 17751-92, http://cdiac.esd.ornl.gov/ndps/alegage.html.

Figure 3-24 The Rising Global Temperature

P. D. Jones, D. E. Parker, T. J. Osborn, and K.R. Briffa, "Global and Hemispheric Temperature Anomalies: Land and Marine Instrumental Records," *Trends: A Compendium of Data on Global Change* (2001), http://cdiac.esd.ornl.gov/trends/temp/jonescru/jones.html.

Figure 3-25 Worldwide Economic Losses from Weather-Related Disasters

Lester R. Brown et al., Worldwatch Institute, *Vital Signs 2000: the Environmental Trends That are Shaping Our Future* (New York : W. W. Norton, 2000).

Figure 3-26 Greenhouse Gases and Global Temperature Over the Past 160,000 Years

J. Jouzel, C. Lorius, J. R. Petit, N. I. Barkov, and V. M. Kotlyakov, "Vostok Isotopic Temperature Record", Trends '93: A Compendium of Data on Global Change (1994), http://cdiac.esd.ornl.gov/ftp/trends93/temp/vostok.593.

C. D. Keeling and T. P. Whorf, "Atmospheric CO_2 Concentrations (ppmv) Derived from *In Situ* Air Samples Collected at Mauna Loa Observatory, Hawaii," *Trends:*

A Compendium of Data on Global Change (August 13, 2001),
http://cdiac.esd.ornl.gov/trends/.

J. M. Barnola, D. Raynaud, C. Lorius, and N. I. Barkov, "Historical Carbon Dioxide
Record from the Vostok Ice Core," *Trends: A Compendium of Data on Global
Change* (1999), http://cdiac.ornl.gov/trends/co2/vostok.htm.

R. G. Prinn et al., "A History of Chemically and Radiatively Important Gases in Air
Deduced from ALE/GAGE/AGAGE" *Journal of Geophysical Research* 115: 17751-
92, http://cdiac.esd.ornl.gov/ndps/alegage.html.

J. Chappellaz, J. M. Barnola, D. Raynaud, C.Lorius, and Y.S. Korotkevich,
"Historical CH$_4$ Record from the Vostok Ice Cores'" *Trends '93: A Compendium
of Data on Global Change* (1994), ftp://cdiac.esd.ornl.gov/pub/trends93/ch4/.

Table 3-3 The Environmental Impact of Population, Affluence, and Technology

Chapter 4

Figure 4-1 Nutrition and Life Expectancy
UN Food and Agriculture Organization FAOSTAT on-line database,
http://apps.fao.org/ (accessed 12/17/01).
World Population Prospects: the 2000 Revision (New York: United Nations, 2001).
http://www.un.org/popin/.

Figure 4-2 Development Costs of New Agricultural Land
Dennis L. Meadows et al., *Dynamics of Growth in a Finite World* (Cambridge, MA:
Wright-Allen Press, 1974).

Figure 4-3 Possible Modes of Approach of a Population to Its Carrying Capacity

Figure 4-4 Feedback Loops Governing Population and Capital Growth

Figure 4-5 Feedback Loops of Population, Capital, Agriculture, and Pollution

Figure 4-6 Feedback Loops of Population, Capital, Services, and Resources

Chapter 5

Chapter 6

Chapter 7

Index

CHELSEA GREEN
PUBLISHING

the politics and practice of sustainable living

Ecology

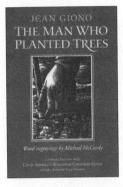

The Man Who Planted Trees
Jean Giono
ISBN 1-890132-32-2
$10.00

CHELSEA GREEN publishes information that helps us lead pleasurable lives on a planet where human activities are in harmony and balance with Nature. Our celebration of the sustainable arts has led us to publish trend-setting books about organic gardening, solar electricity and renewable energy, innovative building techniques, regenerative forestry, local and bioregional democracy, and whole foods. The company's published works, while intensely practical, are also entertaining and inspirational, demonstrating that an ecological approach to life is consistent with producing beautiful, eloquent, and useful books, videos, and audio cassettes.

For more information about Chelsea Green, or to request a free catalog, call toll-free (800) 639-4099, or write to us at P.O. Box 428, White River Junction, Vermont 05001. Visit our Web site at www.chelseagreen.com.

Building

Natural Home Heating:
The Complete Guide to
Renewable Energy Options
Greg Pahl
ISBN 1-931498-22-9
$30.00

Planet

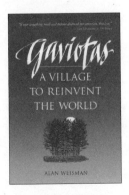

Gaviotas: A Village to
Reinvent the World
Alan Weisman
ISBN 1-890132-28-4
$16.95

Food

The Slow Food Guide to
New York City
Patrick Martins and
Ben Watson
ISBN 1-931498-27-X
$20.00